DEA[illegible]
A LIGHT BULB

John Otten

Blue Ocean Publishing

Death of a Light Bulb

Published by Blue Ocean Publishing
St John's Innovation Centre
Cowley Road
Cambridge CB4 0WS
United Kingdom

www.blueoceanpublishing.biz

Text design and layout by Spitfire Design, Upminster

A catalogue record for this book is available from the British Library.

ISBN 978-1-907527-08-1

First published in the United Kingdom in 2012 by Blue Ocean Publishing.

About the author

John Otten spent most of his working career within the lighting industry, having been educated and trained as a mechanical engineer in the late 1950s; this was at Mullard in Blackburn, a very large factory producing radio and television valves and also having a large wire factory producing filament coils for much of the UK lamp industry. His father, also an electric-lamp maker since the mid 1920s, persuaded him in 1964 to join the family lamp manufacturing business, Kingston Lamp Co. Ltd. in Hull. The business was taken over by Philips Electrical (through Ada Halifax) in 1965. In 1970, he was moved to the UK head office of Philips Lighting, after which he spent the rest of his career in various commercial, marketing and managerial positions. This culminated in him spending nearly eight years in Eindhoven, The Netherlands, in an international role, managing incandescent/halogen product management and development worldwide.

Acknowledgements

Much of the information for this book was researched at the British Library; and many historical publications about the lighting industry were consulted. These included various publications issued in the UK and the USA; these are listed in the bibliography at the end of the book.

Many people in the industry have assisted in completing the information. Of particular note are Ted Glenny from Philips Lighting, and Ian Davies, sadly no longer with us, was Technical Director of Philips Lighting, he helped in many ways, in particular by providing the three large volumes of The History of N V Philips Gloeilampenfabrieken, originally published in Dutch in 1980 but translated into English in 1985. Also, huge thanks to Richard Forster, ex-Thorn Lighting, now a lighting industry journalist, for reading the draft manuscript and giving help and making many suggestions.

A special thank-you must go to a great friend, Michael Grote, Chairman of British Electric Lamps Ltd. This is a very long-established specialist incandescent lamp manufacturing business, dating back to 1920. It is still a significant supplier of specialist lamps to the UK market. Michael not only helped in providing information, but also gallantly encouraged me to finish this book. He contributed many valuable suggestions and also assisted with some of the initial proof-reading.

Contents

1
The Battle to be First

In the beginning God created the heavens and the earth. Now the earth was formless and empty, darkness was over the surface of the deep, and the spirit of God was hovering over the waters.

> *And God said, 'Let there be Light' and there was light. God saw that light was good and he separated light from darkness. God called the light 'day' and the darkness 'night' and there was evening, and there was morning – the first day.*[1]

While it is difficult to imagine a world without day and night, it is also difficult to imagine a world without artificial light. But that is what it must have been like half a million years ago as early man believed the sun travelling daily across the sky was a god, giving light and warmth. The moon at night was a lesser god, shining with a lesser light and supported by innumerable brilliant stars.

Perhaps some 400,000 years ago, fire was discovered, which may have been from a bolt of lightning setting fire to trees or bushes. From this period on, the use of flames was the source of artificial light. The discovery of fire did have a profound effect on mankind, and many early societies constructed myths to commemorate it. Early civilisations used bundles of sticks bound together to make a blazing torch. The first records of primitive lamps date back to 15,000 years ago, found in the Lascaux caves in France; they were made from naturally occurring material, such as shells, horns, rocks and stones. These were typically filled with animal or vegetable fats as fuel and had a fibre wick. The first lighting in Britain dates from 8,000 years ago, when open-wick oil lamps fuelled by fish oil or tallow were common across the whole of Neolithic Europe, as found in Shetland and in the flint mines of Grimes Graves in Norfolk. From this period, there is evidence of ancient lamps being used by many different civilisations, particularly in the Mediterranean region (c.5,000 years ago), in which the fuel would have been plant oils from olives, sesame seeds, nuts, or even fish oil. Wicks would be typically made from plant fibres or bark.

The first candles were most likely made of reeds, which were harvested, peeled and dried, then dipped in tallow and cooled prior to re-dipping up to twenty times to produce a finished candle. These ancient lamps gradually became more sophisticated, with the reservoir and later the flame being enclosed. There is even evidence of the Romans and Greeks using glass spheres filled with water as lenses; true glass lenses were not known until the 13th century. Even the candle can be said to be a relatively recent invention with evidence of the hard animal fat (tallow) or beeswax being used between the 1st and 4th centuries. In the London Museum there are Roman candlesticks from the 2nd century AD. In the 13th century, beef tallow was the cheapest commonly used candle; the more sophisticated candles were made from a blend of beef and mutton tallow.

The process of candle making became much more sophisticated, balancing the wick quality to the thickness of the candle and the melting point of tallow (33°C). Wicks were being made from three or four strands of flax and cotton; the more expensive higher quality candles used by the aristocracy and the Church were made from beeswax. These burned brighter and had less smell and having a higher melting point (68°C) could burn with a finer wick. These candles cost some twenty times that of tallow candles.

During the 18th and 19th centuries, further improvements were made to tallow candles by adding sperm whale oil to the tallow, enabling it to burn with a clear steady flame. In the search for improved compounds to replace tallow, an important by-product was produced. It was the discovery of released glycerine-based compounds, though not eventually used in the candle industry, that led to the foundation of the modern soap industry. Furthermore, it was realised that similar techniques for the separation of energy-rich fatty acids from less volatile compounds could be applied to vegetable oils, particularly coconut oil. This became the founding specialisation of 'Prices' composite candles in the 1840s. By 1851, Prices were producing one hundred tons of candles per week (2.7 million candles) and had their own coconut plantation in Ceylon. In the 19th century came paraffin (wax) and a braided cotton wick, which resulted in the candle we know today. Candle production in Britain peaked in 1917, with an annual production of 45,000 tons to a low of 7000 tons in 1959. Today, Britain produces approximately 14,000 tons annually.[2]

Similarly, the design and manufacture of oil lamps became more sophisticated and along with candles remained the basic form of artificial light until the 18th century. The drive for improvement in artificial light really gathered momentum in the 19th century. Both the oil lamp and the candle had fierce competition from gaslight, which had arrived with the industrial revolution. However, the electrical age was just approaching and it was electricity that drove the early scientists and engineers to seek light from this source of energy.

So from the remotest times until the present day there has been a need to create artificial light. The production of light has been and continues to be one of the most fascinating and exciting kinds of science man has ever tried to master. Like so many inventions from the industrial revolution one invention led to another, and the early developments of the electric lamp were classic examples of this sequence. Electric lamp development has drawn on and continues to draw on a large variety of technologies for its existence and future. Of course, Faraday had to discover electricity to provide the energy source; but then came the need for vacuum techniques, the development of satisfactory glasses, the purification of gases and their behaviour, the refinement of metals, including the very important glass to metal seals, the variety of fluorescent substances. Numerous other engineering feats have contributed to the manufacture and improvement of all lamps known today. Breakthroughs have been achieved by the need for better light sources, whether for longer lifetimes, higher efficiencies, better colour properties, or just lower prices. The two centuries of scientific discoveries, intensive research and development and refinements continually being made every time a light is switched on can never be ignored.

Early electric lamp development continued along three main lines:

Electric arc lamps

Incandescent lamps

Discharge lamps (including fluorescent lamps).

The first notable development was in the early 19th century (two hundred years ago) with the invention of the electric arc by Sir Humphry Davy. In 1808 he gave his famous large-scale demonstration of a continuous luminous arc between two pieces of charcoal to the Royal Institution. The power source for these experiments was huge banks of batteries (two thousand voltaic cells); hence it was not until the discovery by Michael Faraday (1791-1867) of the principles of electro-magnetism in 1831 that gave birth to the electricity that enabled all aspects of the electrical industry to develop.

The first carbon arc lamp was demonstrated in a practical form around 1850; further development took place for the next 40 years to arrive at good working lamps. Davy also demonstrated another way of creating light by passing enough current through a thin piece of wire to make it incandescent. It seems that Davy himself did not actively pursue the development of either source as practical methods of lighting. Perhaps the first permanent installation of the electric arc light was in the South Foreland lighthouse, which was completed in 1858. Various arc lamps continued to be developed. Siemens Brothers & Co. lit the area around the Royal Exchange in the City of London. Other active companies included Brush Arc Lamps and Killingworth Hedges Arc lamps. Indeed in the great Electrical Exhibition at Crystal Palace in 1882, Londoners were for the first time dazzled by the array of electric arc lamps suitable for street lighting.[3] Towards the end of the 19th century this technology then arrived at a natural plateau, giving way to the other lines of development.

The second line of development was incandescent lamps. It had been known from the early 19th century that a heated wire would give off light (filament) when electricity was passed through it. Early inventors struggled to meet the necessary criteria required to make this a viable light source, these being:

- a filament which could be heated to a white-hot temperature and allowed to cool repeatedly without breaking;
- a method of sealing the connections into a container (glass bulb) without the heat cracking or damaging the container;
- methods of preventing oxidation of the wire at these temperatures, i.e. the removal of air (particularly oxygen) by creating a vacuum.

The third line of development was with electric discharges in rarefied gases (this being different to arc lamps). Early experiments were conducted in glass tubes with plugs through which wires passed. A German glass blower, J. Heinrich Geissler (1815-1879) made a significant advance when he sealed platinum wires through soda glass, as they had similar coefficients of expansion.[4] Glassmakers were well aware that most metals did not make good seals through glass. This created the opportunity for scientists to experiment with passing an electric current through different gases and noting that the spectrum of the light varied with the type of gas through which the discharge arc passed. Various papers were published from 1856 onwards; the German physicist, Julius Plücker (1801-1868), published a series of papers between 1858 and 1862; indeed, he liaised with Faraday on the subject.[5] John Hall Gladstone (1827-1902) published a paper in 1860 describing the effect of passing an electric current through a reservoir containing mercury as 'scatters and diffuses it in a vapour with the production of a most intense light'. He analysed the light with the aid of a prism and noted a number of distinct lines of yellow, green and blue parts of the spectrum. There was virtually no red component.[6] It was not until the end of the 19th century that any practical discharge lamps were developed.

As mentioned earlier, it had been known for many years that an electric current would heat up a conductor through which it passes; indeed, many scientists had repeatedly tried to obtain light through incandescence of materials heated this way. Faraday's famous invention of electromagnetism and hence electricity in 1831 continued to lead many inventors in America and Europe to experiment with both the basics of the arc lamp and the far more challenging incandescent lamp. In 1840 W. R. Grove appeared before the Royal Society in London with a demonstration of his batteries; he lit the auditorium with incandescent lamps having a platinum filament. F. de Moleyns obtained a British patent in 1841 for a vacuum lamp with a filament made from a combination of platinum and powdered charcoal. Indeed, an American inventor, J. W. Starr, came to England in 1845 and obtained a patent for two incandescent lamps, one of which had a strip of platinum as the filament and the other a carbonised thread. Perhaps more significant was Starr's description of his construction using a carbonised thread in a torricellian vacuum (made with a column of mercury); Joseph Swan later used this technique. Similarly, on the Continent there were several inventors experimenting along the same lines, all mostly learning from each other. The prize was getting nearer; not only were the inventors aware of this, so also were the industrialists.

Sir Joseph Swan (1828-1914) was a physicist and chemist born in Sunderland; he is generally credited with producing the first practical electric light bulb. At the age of thirteen, he began an apprenticeship with a pharmacist; later he became an assistant in a firm of manufacturing chemists in which he eventually became a partner. This company produced, among other goods and services, photographic plates, for which Swan produced some of his most impressive scientific innovations. In 1862, he patented the first commercially feasible procedure for carbon printing in photography. Later, with the increase in heat sensitivity of silver bromide emulsions, Swan invented the dry plate in 1871, followed by the development of bromide photographic paper in 1879.

It is known that Swan began experiments with incandescence for illumination purposes in the late 1840s and registered a patent for a carbon filament lamp that operated in a partial vacuum in 1860. The most significant feature of Swan's lamp was that it lacked enough residual oxygen in the vacuum tube to ignite the filament (carbonised paper), thus allowing it to glow almost white hot without catching fire. Any incandescent metal or carbon will oxidise or burn if there is oxygen present; but he did not have a good enough vacuum or indeed sufficiently powerful batteries to achieve complete incandescence. Although he obtained a patent in 1864 for his carbonisation process, he knew that further progress, unless he could reduce the presence of oxygen, was not possible, and he turned his attention to other pursuits.

It was through the use of the Sprengel air pump in 1865 that a moderately efficient vacuum could be obtained. In the years to come it was the combination of the Sprengel air pump and improved materials for the filament that set the stage for the inventions to come. From that year onwards development was rapid.

Another key development came in 1870 with the Gramme generator, developed by a Belgian, Zénobe Théophile Gramme; it was the first practical generator for electricity for lighting. This produced a steady source of power used initially for arc lamps in street lighting, which was later adopted by Swan and others for incandescent lamps.

Seeing the effect of the Sprengel vacuum pump, Swan's ambition was revitalised to produce a light bulb, not only with a better vacuum but also an improved carbonised thread as a filament. In 1877, now assisted by Charles H. Stern, he repeated his experiments with carbonised paper and board; but it became clear that the low resistance lamp was not satisfactory and he switched his experiments to high resistance.[7] He realised that he had to solve two important problems: first, early combustion and consequent fracture of the very thin filament, and secondly, the blackening of the bulbs. Realising that both problems occurred together, he assumed they must be related, and finally concluded that the cause of the failure must be attributed to evaporation of the carbon at high temperature under the influence of the high current. At this point Swan must have seen that success was near. On 3rd February 1879, he presented a very important lecture on 'The Electric Light' to the Literary and Philosophical Society of Newcastle upon Tyne. In this paper, he outlined the main developments leading to the arc lamp, developments in electricity generation, materials for lamps and the conviction that the incandescent lamp was the future. He must have known he was near a major breakthrough when he said:[8]

> The principle of lighting by the incandescence of a refractory and badly conducting wire of metal, or by a very thin rod or film of carbon promises to afford the means of avoiding the defects inseparable from the principle if lighting by means of a voltaic arc.

> The principle of incandescence: it appears to be possible to construct a lamp entirely free of machinery and liability to go wrong. A lamp in which the incandescence rod or spiral will not require renewal – in which the light will be perfectly steady and will not dazzle with excess of brilliance – which will not vitiate the atmosphere and can be lighted or extinguished by a touch.
>
> There have been many attempts to realise this idea since J. W. Starr took out a patent (in the name of Edward King) for producing light by the incandescence of continuous metallic or carbon conductors in 1845.
>
> It is evident that both gas light and electric light will have their special uses and that gas will continue to be largely used, perhaps as largely as ever and electric light will also come into extensive use in its own special field.

By now it was clear that this was the main challenge for Swan and with the assistance of some key people around him the breakthrough came later the same year. He was still under the impression he was perfecting inventions made over the previous decades; however, he did start to register his own patents and the culmination was a famous paper presented by Swan in October 1880 to the Institution of Electrical Engineers in London:[9]

> If this idea of carbon were founded in fact, any further attempt to render incandescent carbon lamps durable by means of a vacuum, were mere waste of time, and durable they must be if they are to be of any practical value. Fortunately I did not accept as conclusive the experiments which seemed to show that carbon was volatile, and that the blackening of globes of incandescent carbon lamps was an inevitable result of the

carbon being very highly heated. I know the conditions under which, without exception, all previous experiments had been tried, were such as did not allow to be formed anything approaching a perfect vacuum within the lamp. Screw fittings had invariably been employed to close the mouth of the lamp, and the ordinary air pump to exhaust the air, under such circumstances it was certain that a considerable residuum of air would be contained within it, and also that it would leak. Then, there had never been any thought given to the gas occluded in the carbon itself, and which, when the carbon became hot by the passage of current through it, would be evolved; nor had sufficient care been taken to make the resistance at the points of fixture of the carbon, less than in the carbon to be heated to incandescence. It was evident to me that before any definite conclusion could be arrived at as to the question of the volatility of carbon, the cause of the blackening of the globes and the wearing away of the incandescent rods, we must first try the experiment of heating the carbon to a state of extreme incandescence in a thoroughly good vacuum and under more favourable conditions as to the contact between the incandescent carbon and the conductors supporting it that had hitherto obtained. Accordingly, in October 1877, I sent to Mr. Stern a number of carbons, made from carbonised cardboard, with the request that he get them mounted for me in glass globes by a glass blower, and then exhaust the air as completely as possible. In order to produce a good vacuum it was found necessary to heat the carbon to a very high degree by means of the electric current during the process of exhaustion, so as to expel the gas occluded by the carbon in its cold state, for, otherwise however good the

> vacuum was before the carbon was heated, immediately the current passed and made it white hot, the vacuum was destroyed by the out-rush of the gas pent up in the carbon in its cold state. In order to make a good contact between the carbon and the clips supporting it, the ends of the carbon were thickened, and, in some of the early experiments, electrotyping and hard soldering of the ends of the carbons to platinum was resorted to. The prescribed conditions having been rigorously complied with, it was found, after many troublesome experiments, that when the vacuum within the lamp globe was good, and the contact between the carbon and the conductor which supported it sufficient, there was no blackening of the globes, and no appreciable wasting away of the carbons. Thus was swept away a pernicious error, which, like a lying finger post, proclaiming 'No road this way', tended to bar progress along a good thoroughfare. It only remained to perfect the details of the lamp, to find the best material from which to form the carbon, and fix this material in the lamp in the best manner. These points I think, now satisfactorily settled; and you see the result in the lamp before me on the table.

Swan certainly continued experimenting to improve the lamp by concentrating the vacuum, sealing the lamp and, in particular, the filament. Of note was his preferred construction of the filament; not liking carbonised paper and carbonised parchmentised paper, he eventually discovered that ordinary cotton thread treated in sulphuric acid had the effect of compacting the cotton and making it more uniform. This carbonised parchmentised thread remained Swan's preferred filament for several years. In 1878, 1879 and 1880 he presented many papers and demonstrations in Newcastle, Gateshead, Sunderland and London. These were widely reported in the scientific journals of the day. Furthermore, though he did apply for several patents on related processes, on many aspects it seems he thought he was improving on existing known data. While searching for a better filament, Swan made another significant advance by developing a process for squeezing nitro-cellulose through holes to form fibres; this process was patented in 1883.

Elsewhere in Britain, other engineers and scientists had also been working on similar lamp developments, and forming companies to sell incandescent lamps. George Lane-Fox, having registered unsuccessful patents in 1878, using a high resistance filament of platinum-iridium alloy in an atmosphere of nitrogen and another had another incandescent burner of asbestos impregnated with carbon. In 1880 he did have some success with a lamp using a fibre of French grass treated with hydrocarbon vapour and then carbonised. In addition, he designed an electrical distribution system, meters for measuring electricity used. These patented techniques to make and sell lamps through the Anglo-American Brush Electric Light Corporation. Later Lane-Fox was also to lose his legal battles with Edison & Swan; a similar fate befell Woodhouse & Rawson Lamps. Other companies that came and went in this period were Berstein's Lamps, Crookes Lamps, Crutos Lamps, Gerards Lamps and Pilsen-Joel, mostly losing legal battles to Edison and Swan.

In 1875 Henry Woodward and Matthew Evans in Toronto patented a light bulb but they were unable to raise enough finance to commercialise their invention. An enterprising American scientist, Thomas Edison, (1847-1931), who was considering working on similar ideas, bought the rights to their patent. Capital was not a major problem for Edison; he had the backing of a syndicate of industrialists with $50,000 to invest.[10] In 1876, Edison sold all his Newark manufacturing concerns and moved his assistants to the small village of Menlo Park, 25 miles south west of New York City. He established a new facility containing all the equipment necessary to work on his inventions. This research and development laboratory was considered the first of its kind and was later used as a model for the Bell Laboratories. It is often considered one of Edison's greatest inventions; he registered more than a thousand patents in a variety of fields. Edison was already known for his contribution to the invention of the phonograph and design of telegraph instruments, and at the end of 1877 he turned his attention to what was probably to become his greatest challenge: the development of a practical incandescent electric lamp.

He continued to work on a solution using a lower current, a small-carbonised cotton thread to make the filament and an improved vacuum. Such was the pressure to achieve success that he had a few false alarms during which announcements were made, only to be dashed by lamp failures. Eventually he had a filament that burned for some thirteen and a half hours; this was the first leading up to those that lasted several hundred hours. In November 1879 he submitted his patent to the United States Patent Office and in December held a public demonstration in his Menlo laboratories to demonstrate a light bulb that burned for two days; his patent was granted on 27th January 1880 (Basic Lamp Patent No. 223898).[11] This was to become one of the most controversial patents in the history of the incandescent lamp industry. Edison lodged patents in Canada, France and Great Britain; in the following year he obtained 34 supplementary patents in the USA, all of which concerned the incandescent lamp. A year before Edison registered his patent, he set up 'The Edison Electric Light Company', incorporated in October 1878, and with substantial capital and a well-equipped laboratory. The basic patent was for:

i) a high-resistance filament of carbon in

ii) a chamber made entirely of glass and closed at all points by fusion of glass, which contained

iii) a high vacuum through which

iv) platinum wires passed to carry the current to the filament.

It was a patent based on the combination of known principles, which produced a new product. Edison's success in the field of electric light brought him to new heights of fame and wealth, as electricity spread around the world.

Joseph Swan must surely have regretted not patenting his total system for an electric lamp earlier, as he had patented several processes before Edison applied for his patent for the lamp at the end of 1879, not only in America but also in several other countries, including Great Britain. It is clear that Swan in England and Edison in America had both produced practical lamps in 1879. In a paper given by Swan to the Literary and Philosophical Society of Newcastle in October 1880, it is clear that he was trying to ensure history was recorded correctly:[12]

> You have, of course, all heard that after Mr. Edison abandoned his platinum lamp as impracticable, he invented a new lamp in which carbonised cardboard was used. Here is a diagram of Mr. Edison's carbon lamp, with its horseshoe of carbonised paper.

It is in some respects like mine, latterly I have given up the use of carbonised cardboard, and am now using a material much better, as carbonised cardboard was better than the material previously used. In an article, which appeared in the February (1880) number of Scribner's Magazine, authenticated by a letter from Mr. Edison in the same publication, it is stated that Mr. Edison was the first to use carbonised paper; that, however, is incorrect. And this occurs after the description of the Sprengel pump used in exhausting these lamps: 'Mr. Edison's use of carbon in such a vacuum is entirely new.' Now I daresay there will be many who will remember this little lamp, which I showed here two years ago in action. This lamp has exactly the same simplicity as my present lamp, being composed entirely of three substances, namely glass, platinum and carbon. It is exhausted in exactly the same manner, and to the same degree, as that which Mr. Upton - no doubt in good faith, but still in error -speaks of as *'entirely new'*.

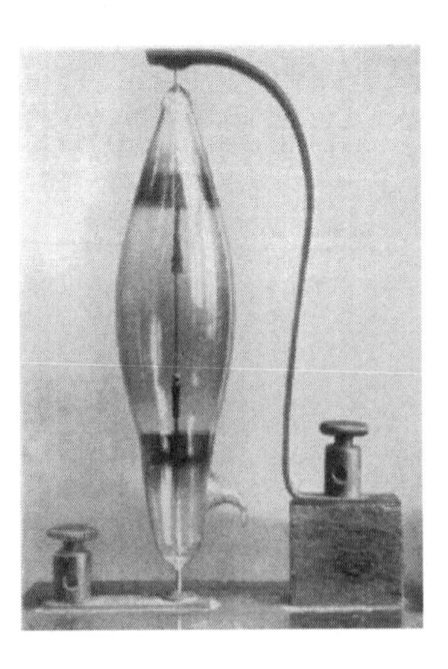

Figure 1: A replica of Swan's first incandescent lamp[13]

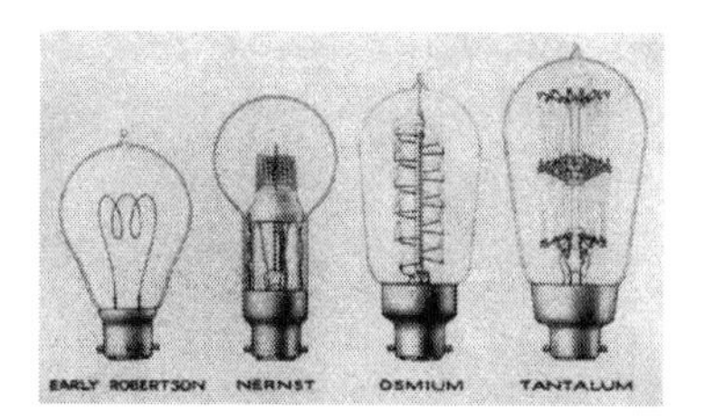

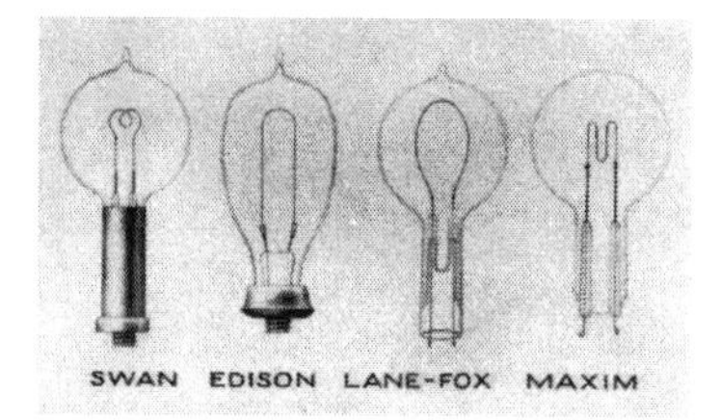

Figure 2: Early examples of incandescent lamps[14]

I do not mention these things in any way to disparage Mr. Edison, for no one can esteem more highly his inventive genius than I do. I merely state these facts because I think it is right to do so in my own interest, and in the interest of true history.

Even after Edison and Swan formed an alliance through their joint company, patent battles would still continue. In 1885 there was a famous lamp patent case of Edison and Swan v. Holland and others. In this case there was an unusual twist in which the infringers were trying to prove 'prior user' by Swan, thereby making the Edison patent invalid, which was one of the cornerstones of the Edison and Swan monopoly. The issue finally went to the Court of Appeal, where the argument eventually centred on the question: was the carbon conductor in Swan's first lamp a filament or not? If it was, then Edison's patent was bad, and the invention would become the property of all. If it was not a filament, then the day would be saved.

All who could be spared in the new Ponders End factory were put on to experimental work, in preparation of the evidence for the case. Numerous specimens were made and assembled before the experts, the lamps were to run to a time trial all night and the results reported to Counsel for evidence the following day. In the course of its deliberations, the Court itself adjourned and moved bodily to the Ponders End factory to see things for itself. The factory was thronged with distinguished lawyers and scientists. After many days of argument, their lordships came to the conclusion that whatever the the carbon conductor in Swan's first successful lamp was, it was not a filament. Edison's patent and consequently their monopoly were upheld, and the beaten competitors left the field. It was a resounding victory that left the company in a strong position, although the decision further obscured Swan's position as prior inventor of the lamp.[15]

In 1897, a German scientist, Walter Nernst, introduced a lamp with a difference, being described as a crossbreed between an incandescent lamp and a gas mantle. The squirted incandescent filament was composed of a mixture of zirconium and several rare earths. This material is a near insulator at low temperature but becomes reasonably conductive at high temperatures. The lamp had to be preheated, originally by a match, but later by a preheating filament of platinum, which was switched off automatically after the lamp became conductive. Because of the pronounced negative temperature coefficient, a current limiting device in the form of a resistor in series with the lamp was necessary, much the same as is done with a modern fluorescent tube. The lamp burned in air, the glass globe only being provided for protection, and this had to be cleaned regularly. The Nernst lamp was expensive and prone to breakdown, but once burning it boasted luminous efficiency two or three times higher than that of the contemporary carbon filament lamp.

In parallel with the Nernst lamp, the first metal filament lamp developed by Carl Auer von Welsbach was introduced, also in

1897. This had an osmium filament, which was made by squirting a mixture of osmium powder and a carbon containing binder. Because of the high price of osmium, burnt-out lamps were much sought after for recycling purposes, a trade-in price even being offered.

In 1902 a Russian chemist, Werner von Bolton, succeeded in drawing metal tantalum into fine wires, and used these to construct a lamp with a tantalum filament. Because of tantalum's low resistance, very long filaments had to be used. A further drawback was that tantalum lamps could only be successfully operated on a d.c. supply, as a.c. made the filament brittle.

None of these lamps, the Nernst, the osmium or the tantalum, survived when the tungsten filament lamp created by Alexander Just and Franz Hanamann in 1903, and improved with the Hans Kuzel tungsten filament, eventually came into production in 1907. This was to create a new standard for the time.

Swan, like Edison, registered a company in 1880, the 'Swan Electric Lamp Company Ltd.', which may have been undercapitalised; in the following year it became the 'Swan Electric Light Company Ltd.' with an authorised capital of £100,000. This new company was better structured to establish lamp manufacturing and to promote the Swan lamp on an international basis. In 1881, the Swan lamp was exhibited at the Paris electricity exhibition and the Swan lighting system was patented in Germany and France as well as in several other countries. Swan himself undertook the post of Technical Director and received large sums of money in consideration for his invention. By this time both the market and the business were developing rapidly; again, a new company was registered in May 1882, the 'Swan United Electric Light Company

Ltd.' with capital of £1m , the factory in Newcastle and a head office in London.[16]

Swan sold his U.S. patent rights in June 1882, to the Brush Electric Company of America. The British Brush Company was the Anglo American Brush Electric Light Corporation; it operated independently as it was established by London financiers in 1880 with an authorised capital of £800,000. This British company, although established to manufacture and sell arc lamps, in 1881 paid £50,000 for the rights to Lane-Fox's incandescent lamp patents. Swan went on to further improve the filament by converting cotton into nitrocellulose, which is soluble in acetic acid. By squirting the solution into a suitable coagulant, he obtained a thin, very uniform thread, which was carbonised to form the filament. This was patented in 1883.

Both these men, being very eminent engineers of their day, will have realised the significance of their discoveries; they were involved with entrepreneurs in setting up companies. An electric lamp suitable for domestic use was an invention waiting to be made; and as already mentioned, many engineers, physicists and inventors, during the second half of the nineteenth century, had viewed this as the 'Holy Grail'. Today both are recognised as having fathered the electric light bulb but Edison was quicker to register his patents. His backers probably had more financial muscle, and established The Edison Electric Light Company in many countries round the world; hence Swan lost much of the credit due to him. However, in Britain, Edison unsuccessfully challenged Swan in the courts for patent infringement; patents were granted for both inventions but declared invalid in every other European country. This forced Edison and Swan to work together and in Britain a company, 'The Edison & Swan United Electric Company', was formed in 1883.

The particular virtue of the incandescent lamp was clearly its suitability for small-scale uses. Although the much-celebrated Holborn Viaduct scheme in 1882 has been regarded by some as the first street lighting scheme in which the Edison Company also supplied electricity to private users along the route, there was an earlier lesser-known scheme in Godalming, Surrey. In 1881, the Godalming Town Council was able to derive power from a waterfall on the River Wey and entered into a contract with Siemens to provide street lighting and some private consumers with electricity. Authority to dig up the streets for cables did not exist and therefore they were laid in the gutter.[17] Whilst this was a brave attempt by Godalming Council, it was sadly premature and too few private consumers applied; the scheme only lasted three years before the street lighting went back to gas. Nevertheless it achieved the reputation of the first electric street lighting in the country.

Although still in its infancy, the incandescent lamp was seen by many as a competitor for gas lighting and the now well-known and established arc lamp. Indeed, one of the first football matches to be played under artificial light was probably in 1878, when 20,000 people went to see a floodlit match at Sheffield, where powerful arc lamps powered by a Siemens dynamo were used.[18] Sir William Siemens was already a prominent businessman of the time, managing Siemens Brothers, a subsidiary of Siemens & Halske of Germany; both companies were to play a major part in shaping the lamp industry of the future.

While many scientists and inventors from Europe and America

Figure 3: The town of Godalming illuminated by electric light 21st November 1881[19]

played a part in the development of the electric lamp, the two key figures were Joseph Swan and Thomas Edison. Within Britain, Swan was recognised by being elected to the Royal Society in 1884, made president of the Institution of Electrical Engineers from 1898/99 and president of the Society of the Chemical Industry in 1901. He was awarded an honorary doctorate from Durham University and knighted in 1904.

This was the start of one of the great industries of our time , and certainly one of the first to be formed on the patents and patent battles, licence and cross-licence agreements which were to become a major feature in the shaping of the industry in the next century.

Some Key Incandescent Lamp Developments up to 1878[20]

Year	Researcher	Country	Filament	Atmosphere	Enclosure
1838	Jobard	Belgium	Carbon rod	Vacuum	Glass
1840	Grove	England	Platinum helix	Air	Inverted glass beaker in a dish of water
1841	De Moleyns	England	Powdered charcoal between two platinum helices		
1845	Starr	USA	(1) Platinum strip	Air	Glass sphere with stopper
			(2) Graphite stick	Vacuum	Glass sphere above column of mercury
1848	Staite	England	Platinum with iridium rods	Air	Glass sphere with stopper
1850	Shepard	USA	Charcoal cylinder against sphere of the same material	Vacuum	Glass sphere with stopper
1852	Roberts	England	Graphite stick	Vacuum	Glass sphere with stopper
1854	Gobel	Germany	Carbon wire (?)	Vacuum	Glass sphere with stopper
1858	De Changy	Belgium	(1) Platinum helix	Vacuum	Hermetically sealed glass sphere
			(2) Carbon rod		
1860	Swan	England	Strips and helices of carbonised paper and cardboard	Vacuum	Jar or bell on copper base plate
1872	Lodyguine	Russia	(1) Carbon rod	Vacuum	Hermetically sealed glass sphere
			(2) V-shaped piece of graphite	Nitrogen	Glass sphere with stopper
1873	De Khotinsky	Russia	Carbon rod	Vacuum	Hermetically sealed glass sphere
1875	Kosloff	Russia	Graphite rod	Nitrogen	Hermetically sealed glass sphere
1876	Bouliguine	Russia	Graphite rod	Vacuum	Hermetically sealed glass sphere
1878	Edison	USA	Platinum helix	Air	Glass sphere
1878	Lane-Fox	England	Platinum iridium	Nitrogen	Glass sphere
			Asbestos carbon		

Note: * The principal types of incandescent lamps developed in the experimental period are listed above.
* The second stage of the incandescent lamp development commenced about 1878 and culminated in the commercial lamps of 1880 and the following years.

[1] The Bible, Genesis chapter 1 v.1-5
[2] Paper 'Early Lighting' by Roy Wilde (Clerk to the Tallow Chandlers Company)
[3] From the Beginning by Henry Loring 1905
[4] Lengthening the Day by Brian Bowers 1998
[5] Faraday – Plücker correspondence in the archives of the Institution of Electrical Engineers, collection SC MSS 2, D.J.Blaikey papers
[6] 'On the electric Light of Mercury' an article by J.H.Gladstone in Philosophical Magazine
[7] The Electric lamp Industry: Technological Change and Economic Development from 1800 to 1947 by A. A.Bright 1949
[8] Paper presented by Joseph Swan to the Literary and Philosophical Society of Newcastle upon Tyne on 3.2.1879
[9] Paper presented by Joseph Swan to the Institution of Electrical Engineers in London, from The Engineer, 29.10.1880
[10] History of Canadian Invention by Dr. A.J.Carty, 1999, National Research Council
[11] The Electric lamp Industry: Technological Change and Economic Development from 1800 to 1947 by A.A.Bright,1949
[12] 'Electric Lighting' A lecture by J. W. Swan on 20th October 1880 to the Literary & Philosophical Society of Newcastle
[13] Illustration from 'Sir Joseph Swan FRS' by A. Memoir
[14] Illustrations from Marconi-GEC archives
[15] The Pageant of the Lamp by the General Advertising Company of London Ltd.
[16] The History of N.V.Philips Gloeilampenfabrieken by A.Heerding, vol.1, 1980, translated by D.S.Jordan in 1985.
[17] The 'Graphic Magazine' 12th November 1881
[18] 'Sheffield Daily Telegraph', 15th October 1878
[19] Illustration taken from 'The Graphic' journal 21st November 1881
[20] The History of N.V.Philips Gloeilampenfabrieken by A.Heerding, vol.1, 1980, translated by D.S.Jordan in 1985.

2
The Emergence of Winners

Joseph Swan and Thomas Edison were only two of a long line of scientists to pursue an electric incandescent lamp, but their timing was right and they, plus the world around them and in particular the financial institutions of the day, were aware of the importance of the patents. The possible manufacture on a commercial scale excited businessmen and financial institutions. There was a huge and growing need for artificial light and commercialisation of this invention would follow rapidly.

The Edison and Swan United Electric Company formed in 1883 traded as Ediswan, and this company with the Edison Electric Light Company in America defended their patents aggressively, successfully suing many other lamp makers. However Swan sold his American patent rights for the 'squirted' filament to the Brush Electric Light Company. Charles F. Brush was an American inventor who wanted to put Cleveland on the lighting map in 1879; he had developed an arc lamp. The first permanent street lighting in America was installed in that year after its demonstration. Following the success of Edison's incandescent lamp, also in 1879, Brush realised they too had to get into the incandescent lamp business if they were to survive. In that year Brush sent Thomas Montgomery to England in an effort to market his arc lamp. This led to the formation of the Anglo-American Brush Electric Light Company; the plan was to make both arc and incandescent lamps in England. However, this led to a lengthy legal battle in Britain, resulting in the Anglo-American Brush Electric Light Company giving up filament lamp manufacture, and Edison and Swan United Electric paying royalties to use other patents owned by the the Brush Company. In 1882, Thomas Montgomery returned to Cleveland to manage the newly formed Swan Incandescent Electric Light Company of New York and to market the Swan lamp system in America. In 1883, the Brush Electric Company acquired the American rights to the Lane-Fox incandescent lamp patents and added these lamps to their range. The Lane-Fox lamp proved to be unsatisfactory, and in 1884 the Brush Company tested Swan lamps from England. Subsequently, in 1885, they purchased the Swan Lamp Manufacturing Company in Cleveland and the rights to manufacture and sell the Swan lamp design in America. The Swan Lamp Manufacturing Company was eventually dissolved in 1895 and the lamp activities of the Brush Electric Company in America became part of the National Electric Company.

Swan and Edison never actually met, a fact that indicates the hold their industrial and financial backers had over their companies; both were scientists contributing huge advancements to science. Edison was the most outward in promoting his inventions, as he also had a keen business brain; and he had the ear of leading industrialists in America. He held regular meetings with the press, extolling his laboratory facilities, and his ability to beat other inventors. Indeed, it was said that Edison's pronouncements made thrilling reading for a public that was agog with the elusive possibilities of electricity. Twice he announced he had discovered the electric lamp, only to later find he had failed. When the public were becoming sceptical, suddenly, on 21st December 1879, the New York Herald carried a major article headed 'Edison's Light – The Great Inventor's Triumph is Electric Illumination'. A few days later he demonstrated some lamps outside his laboratory. He had applied for patents in America, Britain and many other countries in November 1879. The American patent was issued in January 1880. Edison had a good track record on inventions and a high public profile, which helped him considerably in raising funds for further research and developing production. This not only made him a great fortune but also was to be the basis for what was to emerge some years later as General Electric of America.

Substantial profits were made on the monopoly rights given to the Edison Electric Light Company under its American patent. Consumers of the day had been waiting for this invention; many scientists had spent years trying to produce a successful electric lamp, and the company aggressively defended its patent. Whilst competitors sprung up all over the world, following Edison, under competing patents, the Edison Electric Light Company expanded very rapidly. It was reported that there were over 700 Edison factories by 1886. However, in the period 1885-1901 the Edison companies spent $2 million defending their patents, nearly always successfully. Many smaller companies could not afford the legal action(s) and this often led to liquidation or forced sale.

In 1889, several American lamp companies were consolidated into the 'Edison General Electric Company' in which the majority of the $12 million capital was controlled by the influential financier from New York, Henry Villard, and the great German industrialist Werner Siemens of Siemens & Halske, who died three years later. Thomas Edison and his associates had built up a conglomerate of companies, which had experienced tremendous expansion since 1879, and there was urgent need to consolidate and reorganise. However, the Edison Electric Light Company retained its separate corporate identity for many years after consolidation.[1]

It was not until 1891 that the question of the monopoly of incandescent lamps arose in the United States. By then, the Edison Electric Light Company had become a monopoly and part of the new enterprise. However, there remained two major competitors in America, Thompson-Houston and Westinghouse; and in 1892, with the original patents due to run out in 1893, there was a merger early that year with the Thompson-Houston Company to form the 'General Electric Company', known today as GE of America. The dynamic founder of Thompson-Houston, Charles Coffin became the first president of this huge new enterprise; indeed, many of the key positions went to the Coffin men. The relative position prior to the merger can be seen in the table below.[2]

	Edison General Electric	*Thompson-Houston Electric*
Shareholders' Equity	$15,000,000	$10,000,000
Turnover	10,000,000	10,300,000
Profit	2,100,000	2,700,000
Number of employees	6,000	4,000

The monopoly of the American lamp market was complete and remains so to the present day. GE of America has a wide portfolio of electrical products dominating the American market, ranging from the creation of electrical power stations to aircraft engines and in finance; GE Capital is one of the main cash generators.

On the European Continent, developments were also very rapid; in Germany, Werner Siemens took the lead with his own development in 1880 and production of incandescent lamps in 1882. In 1881, Edison exhibited his lamps in a Paris exhibition and raised considerable capital, enabling him to set up two companies: Compagnie Continentale Edison, to exploit his patents, and the

Société Industrielle et Commerciale as a manufacturer. Germany had been freed from patents, which led to the industry progressing more rapidly, not only in volume output but also in manufacturing techniques. Many new companies entered the field of electric incandescent lamp manufacture. All this eventually led to a combination of licensing agreements or product diversification to get around the patents.

The Edison Electric Light Company was established in many countries throughout Europe, and it had a significant influence by establishing licensing agreements for the exploitation of their patents. By 1890 a young Dutchman, Gerard Philips (1858-1942), studied engineering initially at Delft and worked in shipbuilding in Glasgow and then in London, where he joined the Anglo-American Brush Electric Light Corporation as an engineer. This was clearly a significant influence in generating his interest in electric lamp innovations. In fact, in London in 1889 he had closely watched a patent case brought by Edison and Swan against the Anglo-American Brush Electric Light Corporation which Edison & Swan won and agreed to purchase the Brush patents. Hence Brush ceased to manufacture lamps in Britain. In the Netherlands there was an absence of patent legislation since the suspension of the Patent Act in 1867, and extensive discussions continued on the advantages and

Gerard Philips 1858-1942

disadvantages of patent legislation. Nevertheless, this offered an opportunity to Gerard Philips when he returned, initially, to Amsterdam and met up with Jan Jacob Reesse (1883-1910), a chemical technologist. The two men teamed up to develop an incandescent lamp and in particular the filament development. They also had the help of Charles John Robertson (1860-1909) from London, who was working in the Netherlands at that time and helped in the cost estimates for the construction of a factory to produce 500 lamps a day. Then with the support of Gerard's father Fredrick (1830-1900), a site was found in the southern town of Eindhoven, where they (Gerard and his father Fredrick), formed Philips & Co. in 1891 to produce carbon filament lamps. It would appear that C. J. Robertson wanted a partnership but this did not fit into Gerard's plans. He wrote to him saying 'We think the business is too small to allow three partners'.[3] While the original plan may have been to produce 1,000 to 2,500 lamps a day, they opted to start with a pilot plant producing 500 lamps a day. Competition soon grew in the period up to 1893, when prices began to fall sharply, and the quality of the lamps deteriorated as many more manufacturers entered the market as the initial patents ran out. In 1892/3, Gerard Philips was producing significant volume, although like many companies of the day they found that prices fell by 50/60% in less than two years and they

Anton Philips 1874-1951

were not able to make a profit. Once the factory was established, it was clear Gerard needed to spend his time on development and production. Fredrick recalled his younger son Anton (1874-1951) from London where he was training as a stockbroker and in January 1894 joined his brother, taking responsibility for sales; he quickly developed a shrewd commercial brain. The brothers realised that the only way to get back into profitability was to produce in volume. This was the start of an internal competition between the brothers where Gerard tried to produce more than Anton could sell, and vice versa. This was the start of the great family partnership that was to last for nearly fifty years. Anton developed a very shrewd commercial brain and became the first President of N. V. Philips Gloeilampenfabrieken.

C. J. Robertson stayed with Gerard Philips for some time but was more ambitious. He was the son of a London bullion dealer, educated at Merchant Taylors' School in London and won a scholarship to St. John's College, Oxford that he was unable to take up. He attended Finsbury Technical College to study electrical engineering and analytical chemistry. In 1881 he joined the Anglo-American Brush Electric Light Corporation and became an assistant to George Lane-Fox. Later, he accepted the post of Manager of the incandescent lamp department of Pilsen-Joel; in 1885 he joined Alexander Berstein, for whom he equipped a new lamp factory in London. Like so many other small lamp manufacturers, they closed under the threat of legal action over patents. For a short period he sold lamps privately to former Berstein clients in Germany and France. With the lamp industry still very much in its infancy in the Netherlands, this is where found his knowledge was of value. After the period with Philips, he realised that many of the Edison patents would lapse in

England in 1893, and that the lamp-making industry, also in its infancy, would be capable of giving employment to many thousands of workers and could be as successfully carried on in England as abroad. Of course others had the same idea and when he returned to England in 1894 he found the field fully prepared for him.[4] His first major assignment, back in England, was to assist in starting the incandescent lamp factory for GEC. This was to be in the buildings in Brook Green, London, which had been occupied by the Anglo-American Brush Electric Light Corporation which had lost its patent battle with the Edison & Swan monopoly.

Sir William Siemens. Born as Carl Wilhelm Siemens 1823-1883

Another company that was to have a major influence on the lamp market not just in Britain but the world over, was Siemens. The original British company was founded in 1858 and was then part of the German firm of Siemens & Halske. William Siemens, in 1843 aged 19, came to London from Germany. He sold his first patent for the then huge sum of £1,600, and instead of returning to Germany, decided to stay. Later, in conjunction with his elder brother Werner, he took out a string of patents. These were unrelated to the electric lamp industry and were mainly in telegraph equipment technology. In the meantime, Werner Siemens joined with Johann Halske, a young mechanic, and founded the company Siemens & Halske in Germany in 1847. In 1858 Siemens, Halske & Co. was founded in England with William, Werner and Halske as partners; however, Halske's role only continued until 1865 when, after a disagreement, the Siemens brothers took over the assets of Siemens, Halske & Co. in London. Now William and Werner were partners, with William as head of the new company , taking half the profits. By 1870, the share capital of Siemens Brothers was Werner 40%, William 30% and a third brother Carl 30%.[5] The 'House of Siemens' was by now firmly established, not only in England, where William was knighted, but also in several other European countries, with the lead and developments coming from Germany. Sir William Siemens became one of the leading businessmen of his time in the emerging electrical industry. When a Standing Committee investigated 'Lighting by Electricity' in 1879, Sir William Siemens was one of several distinguished witnesses advocating electric light for public places.[6] In 1880, Siemens Brothers registered as Siemens Brothers & Co. Ltd.

At that time, the arc lamp was the main form of electric light; it was necessary to get away from the few large arc lamps to finding some means of providing a small number of independent lamps, this being the great electrical problem of that time. Already shareholders in gas companies were starting to panic and Sir William Siemens, now being regarded by the public as an oracle in these matters, tried to calm these fears through a letter to 'The Times',[7] stating that there would still be plenty of work for gas to do. This did have a stabilising effect on gas shares. Of course it was the eve of the invention of the carbon filament lamp. After Sir William Siemens died in 1883, Alexander Siemens provided the leadership within the company and of electric lamp development. While Alexander soon had little time for research, he had as an assistant Dr. Eugene Obach, the firm's chemist, who spent much of his time in his laboratory working on improvements to the carbon filament lamp. Indeed it was probably through his work in removing any remaining air in the bulb to improve performance that led Langmuir of GE, many years later, to use argon in the incandescent lamp.[8] When Obach died in 1899, and with him much of the research being carried out in Woolwich, Siemens Brothers had to rely almost entirely on the research being carried out by their sister company in Berlin.

Siemens & Halske had invested considerable research into the incandescent lamp; they became a major producer in Germany with significant sales in Britain through Siemens Brothers. Indeed, in their search for a substitute material for carbon, they experimented with several rare metals such as vanadium and niobium, but the melting point was too low. Finally, their researcher Dr. von Bolton obtained a minute piece of metallic tantalum, which proved both tough and malleable. It was from this discovery that he went on to obtain pure metallic tantalum, which was capable of being drawn into filaments. So tantalum proved to be the first successful substitute for the carbon lamp, being more robust, and consuming approximately half the current for the equivalent light. A successful patent was registered in Germany and Britain. Production in Germany commenced in 1905-6, but it was not until 1908 that Siemens Brothers established a factory in London; with large-scale advertising, they gained a significant market share. Even while this was going on, competitors were still working on other materials. Tungsten had been discovered by Hannaman and Just, so the tantalum lamp was relatively short lived. While tantalum was more robust, tungsten was more efficient. In 1911, their production was converted to the more efficient tungsten lamp, and now they were established as one of the principle manufacturers in Britain along with GEC and

British Thompson Houston. This was the position until the outbreak of war in 1914, when the Government took over the assets and all contact with Siemens & Halske in Germany was severed. It was at this time that they started their own research and development programme.

In 1894, the two leading producers in Germany, Siemens & Halske and Allgemeine Elektricitäts Gesellschaft (A.E.G.), met with the objective of raising standards and stabilising prices; it could be regarded as one of the earliest lamp rings. This was the precursor to a much more comprehensive and international agreement 'Verkaufstelle Vereinigter Glühlampenfabrieken' (Association of United Incandescent Lamp Manufacturers) which was signed in Berlin in 1903. The participants were the main producers in Germany, Austria, Hungary, Holland, Switzerland and Italy.[9] GE in America was also active in 1896 in setting up the Incandescent Lamp Manufacturers Association, which consisted of GE and six other manufacturers, previously bitterly competing companies. Indeed, after its formation, another ten joined, and the Association's objectives were to fix lamp prices, and allocate business and customers to each.

Later in the 1890s, AEG, Siemens & Halske and Philips began to dominate the continental market. In spite of fierce price competition, the gas industry tried to respond with the development of the 'gas mantle', which gave light at a lower cost than the carbon filament lamp. This revival was short lived given the greater convenience of electric lamps, the developments yet to come and the rapidly growing use of electricity, which was spreading rapidly. Many of the same companies, for example Edison (as GE), were very active in promoting uses for electricity.

Most governments were supporting the introduction of generating stations, so the demand for electrical equipment of all kinds grew steadily.

In Britain, several engineer/scientists had also been working on similar lamp developments, for companies selling incandescent lamps. George Lane Fox, having registered unsuccessful patents in 1878, in 1880 made a lamp using a fibre of French grass treated with hydrocarbon vapour and then carbonised. This was more successful; he again patented it, and subsequently used some of the patented techniques to make and sell lamps through British Electric Light Company. However, they lost their legal battle with Edison & Swan, and a similar fate befell Woodhouse & Rawson Lamps of London when in 1884 legal action was taken against them. The case against Woodhouse & Rawson was based on Edison's basic patent on the carbonised filament and the flashing process devised by Sawyer and Man, referred to as the 'Cheeseborough Patent' after the patent agent F. J. Cheeseborough.[10] This patent had been purchased by Edison & Swan; its validity in Britain was not taken seriously at first. The claim to the exclusive rights in respect of the flashing process was regarded as no more than a scientific experiment relating to an impractical lamp dating from 1878, and not seen by lamp manufacturers as a threat. In 1886, after what was regarded as an extremely weak defence – 'the case was almost a *reductio ad absurdum* of the expert farce'[11] – in spite of widespread criticism, the court found in favour of Edison & Swan. After some hesitation, in 1887 the Court of Appeal upheld Edison's basic patent. The consequence was that nearly every manufacturer in Britain making high resistance lamps with a carbonised filament, implying the use of the flashing process, was confronted with the choice between the production of incandescent lamps with their considerable attendant risks, or transferring their activities abroad. Indeed, this did result in some moving production abroad and many of the younger talented engineers moving to continental Europe. By 1889, the Edison & Swan United Electric Co., after many successful lawsuits, had succeeded in ridding itself of almost all domestic competitors. This was probably a major contributory factor to the temporary stagnation in the development of this young and vital industry within Britain, but doubtless, the Edison & Swan Company was perceived as a clear winner of the initial period of the lamp industry.

Companies that came and went in this period were Crookes Lamps, Crutos Lamps, and Gerards Lamps. Others, like Alexander Bernstein, did not give up, but transferred his production from London to Hamburg in 1888. Another great loss to the British lamp industry was that of Charles Stearn (1844-1919), who was Swan's partner; he went to Switzerland to set up as a lamp manufacturer, where incandescent lamps were not protected by patents. Previously, Stearn had been managing the Swan United Electric Light Company's factory in Kalk, near Cologne in Germany. The following note to his closest staff communicated this momentous decision for Stearn:

> *I would take this opportunity to inform you that I have resigned my position as director of the Swan Company, and that from the 31st instant I shall no longer be connected with the company in either [a] technical or commercial capacity. As the Swan lamp was a joint invention on the part of Mr. J. W. Swan and myself, and as I have been in charge of its manufacture from our earliest tests in 1877 until now, I shall*

watch its further development with interest, and it is my hope that the good reputation which it has so far enjoyed will be maintained in the future.[12]

Following this, in February 1890 Stearn set up the Zürich Incandescent Lamp Company, bringing with him several engineers from Britain and also training many Swiss engineers. He not only formed a significant factory but also laid the foundation of the Swiss lamp industry.

Indeed it was with some disappointment that the City of London saw so many lamp companies either disappear or leave to produce lamps elsewhere in Europe. One of the last hopes was the Anglo-American Brush Electric Light Corporation, which announced in July 1889 that it was to cease production of lamps in the United Kingdom. The company had reluctantly agreed to purchase lamps from Edison & Swan on the most favourable terms, which the latter granted to their dealers.[13] In exchange Edison & Swan had accepted a licence from Brush on the Lane-Fox patents, for which they would pay royalties of a farthing per lamp,[14] but in fact made little or no use of the patents. The protracted legal battle on the subject of patents, which became known as 'The Great Incandescent Lamp Case', ended after three years in a compromise. Brush had been obliged to indemnify their agents and clients against the consequences of legal action threatened by Edison & Swan. Proceedings had been instituted against the Jablochkoff Electric Company and the General Electric Company, both being Brush agents, and against their client, William Holland, manager of the Albert Palace. This was an indirect attack on the Brush position; the court ruled that they could retain their stock of lamps but could not add to it. Brush felt they had no alternative but to attempt to undermine the Edison & Swan position in the area of patents. The Chairman of Brush, Lord Thurlow, felt they had not only a good case but also an overwhelmingly strong case, and expressed his distaste for the affair in this statement to the shareholders:

> This litigation, while exceedingly to be regretted, from its serious character to your interests, is however, of far graver importance to the Edison & Swan Co., involving as it does the validity of the most important of their lamp patents, which as would appear from their Balance Sheet, stand at about £300,000, and upon which they rest their claims for an absolute monopoly on lamp making.
>
> The Board takes this opportunity, to place on record their regret that their efforts to bring about an amicable settlement, consistent with your interests, have failed, and that everything points at present to the necessity of the case being submitted to the final decisions of the Courts of Law.[15]

This case against Edison & Swan came before the court in April 1888 and lasted 23 days. Mr Justice Kay found himself confronted by a maze of technical details and statements, often contradictory, presented by many expert witnesses including Swan himself. He proposed that each of the parties, in the presence of the other and of an independent expert, should have a number of saleable lamps, manufactured by an 'intelligent workman' and in accordance with the specifications set out in the Edison's basic patent.[16] The objective of the test was to establish whether, on the basis of the Edison's description, a successful lamp could have been manufactured other than by a person such as Lane-Fox, who already possessed as much or more knowledge. The president of the Royal Society, Professor G. G. Stokes, was invited to act as an independent expert and report the findings to the court. Two places were chosen to carry out these tests, one being the Edison & Swan works at Ponders End and the other at William Crookes' laboratory in Kensington Park Gardens; the tests took place in June 1888. Crookes and Professor Silvanus Thompson acted as observers representing Brush, and Professor James Deware and Dr. John Hopkinson acted on behalf of Edison and Swan. The results of these tests showed that at the Edison & Swan factory:

- 2 lamps failed immediately
- 17 others failed after an average of 13 hours
- 12 continued to burn after 32 hours

and at the Crookes laboratory

- 4 only burnt for more than 60 hours (out of 100 lamps)

Hence the specifications of 1879 could not be said to be a commercial success, defined as lamps that could be made in large quantities, cheaply and above all, possessing a long life. The conclusion was that the earliest incandescent lamp manufactured commercially by Edison, with the bamboo filament, patented in September 1880 and introduced at the Paris Electricity Exhibition in 1881, could not have been a commercial success. Mr Justice Kay ruled in advance that the claim that the basic patent must in principle be held to apply to all types of carbon filament, including bamboo, was too far-reaching. Based on the evidence of the tests, information published in 1879, no researcher could have produced a successful lamp unless like Edison, Lane-Fox, Swan and others acquired specialist knowledge. Mr Justice Kay's verdict delivered on 16th July 1888 therefore destroyed the validity in Britain of the Edison's basic patent on the following grounds:

- The claim for a monopoly of incandescent lamps containing a filament of carbon for a burner must be regarded as far too wide considering how little Edison had actually invented.
- The specification did not describe a lamp which ever became, or could have become, commercially successful.
- The directions contained therein were so insufficient that no one could have made the carbons described without considerable previous experiment.
- One of the processes described, namely mixing the carbons with volatile powder, was partially injurious if done as Edison directed.
- Coating with non-conducting, non-carbonising substance, if not injurious, was of no practical utility.
- The same could be said of coiling the filaments, on which the patentee laid great stress.[17]

While in the earlier case against Woodhouse & Rawson, the Court of Appeal had upheld Edison's basic patent, though only after hesitation, now, after using mainly technical arguments, this proved decisive. The court had upheld the other minor issue in dispute concerning the flashing process (the Cheeseborough patent bought by Edison & Swan), as previously outlined.

Clearly, electrical engineers in Britain and in Germany welcomed the outcome of this case and concluded that there was no obstacle to the sale of the unflashed Seel lamps in Britain. This was a process developed by Carl Seel of Charlottenburg, Sweden, which embodied a method of making a homogeneous filament without resort to the flashing process. Seel's filaments were made from silk or wool impregnated with a mixture of sodium silicate and gum arabic.[18] Woodhouse & Rawson reopened their lamp works in London and announced an order for 200,000 lamps, a new type, which did not require to be flashed.[19] Within days of Mr Justice Kay's verdict the General Electric Co. (GEC) approached Johan Boudewijnse of Middelburg, Holland, for a contract to supply unflashed lamps. Unfortunately, they were unable to supply at that time.[20] The Anglo-American Brush Light Corporation felt it had emerged the victor and went to great lengths to point out that its 'Victoria' lamp in no way infringed any patent. Indeed, they commenced building a new factory at Brook Green in London.

However, this was all a little premature, as Edison & Swan engaged new legal advisers and entered an appeal against the judgment, resulting in the case being reopened in December 1888. Again Edison & Swan pursued not Brush, but their clients. The Court of

Figure 4: Advertisement for the Steel lamp

Appeal gave the verdict on 17th February 1889[21] and, primarily on legal grounds, revalidated Edison's basic patent. This, however, did not signify the end of the conflict, which had dragged on for three years, as a new trial would be needed to establish whether the Brush lamp infringed the now restored Edison patent. In the short term, neither party had any chance of succeeding; indeed, it was said that the patent regime in Britain was 'akin to the state lottery in France or Italy'.[22] The situation was made even more intractable when Brush announced its intention to lay the most recent verdict and, by implication, the question of sound patent legislation, before the House of Lords. The outcome would have been predictable; and Brush announced to its shareholders that incandescent lamps only represented ten per cent of their business, stating they could continue without making lamps.[23] In the compromise reached with Edison & Swan, Brush undertook not to produce lamps in Britain while Edison's basic patent remained in force. In May 1889, Ediswan & Swan announced their intention to terminate their installation business and concentrate on the manufacture of lamps and fittings.

Later, at an Extraordinary General Meeting in July 1889, the Board of Brush proposed to change the name of the company from Anglo-American Brush Electric Light Corporation to Brush Electrical Engineering Company. The Board also announced that it had bought a locomotive and carriage works in Loughborough, taking them away from lamp manufacture. Some years later, in 1893, the new and unused Brush factory at Brook Green in London was sold to the General Electric Company (GEC).[24]

In 1888 the Electric Amendment Act was passed, giving rise to an upsurge of electrification and a dramatic increase in the installation of lamps which assured the immediate future of Edison & Swan until its basic patent expired on 10th November 1893. From this date onwards the company would come under severe competition, not only at home but also particularly abroad.

Figure 5: A GEC 1890 catalogue cover

Of less significance at the time was the arrival, in 1880, in Britain of Hugo Hirsch, from Munich, Bavaria, who later changed his name to Hirst and took British nationality. He initially worked for the Electrical Power Storage Company but after they got into financial difficulties he went to work for another Bavarian immigrant, Gustav Byng (originally Binswanger), who, in 1880, had started a small business selling components for the infant electrical industry in Great St. Thomas Apostle in London. Hugo Hirst joined Gustav Byng in 1883 and later in 1885 asked Hirst to set up a department to sell electric light accessories. The original company G. Binswanger & Co was not incorporated and by 1886 it changed its name to the General Electric Apparatus Company (G. Binswanger). Business grew rapidly and in 1889, the General Electric Apparatus Company was incorporated as a private limited company under the name the General Electric Company with a capital of £60,000. Gustav Byng became its first Chairman. Authors on the history of G.E.C. later claimed that this was the first company to have 'General Electric' in its title. Hirst initially resigned but was later reinstated with 20% of the shares, creating a strong position for him in the newly incorporated company. In the same year they moved their premises to Queen Victoria Street, London.

Edison & Swan had defended their original patents aggressively until initial patents ran out in 1893, but these patent battles were set to continue. On the lapse of the original patents, imports began to increase significantly, often at prices which the British manufacturers had difficulty in competing with. It was in that year that G.E.C. decided to invest in the manufacture of lamps, and Hirst had visited lamp factories in Germany, Austria and America and then teamed up with C. J. Robertson, who had started and run several small lamp factories. Providing half the capital to set up Robertson Electric Lamps Ltd., they purchased the new unused Bush factory at Brook Green. Under the watchful eye of Hirst and the strong support of the G.E.C. sales force, Robertson Electric Lamps made good progress as a manufacturer of carbon filament lamps.

The electric incandescent lamp industry continued to grow rapidly but became very competitive. In 1901 a great German industrial crisis resulted in many factories having to go on short time working, putting huge pressure on the lamp industry. Philips, being

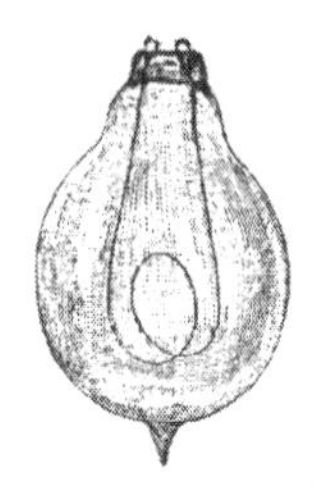

1. The "Robertson" Incandescent Lamp fitted with bottom loop.

2. The "Robertson" Incandescent Lamp fitted with Edison screw cap.

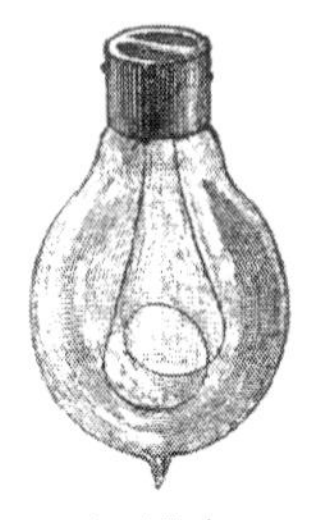

3. The "Robertson" Incandescent Lamp fitted with brass cap.

Figure 6: Examples of the early Robertson lamps

very volume orientated, kept their factories open by selling large volumes of lamps, at very low prices, through agents into France, Spain and Italy. This more than compensated for the loss of a market in Germany, but since other companies also tried the same tactics throughout Europe, this eventually led to the formation of an association of carbon filament lamp manufacturers. The international agreement *'Verkaufstelle Vereinigter Glühlampenfabrieken'* (Association of United Incandescent Lamp Manufacturers) was signed in Berlin, 1903, by the main producers in Germany, Austria, Hungary, Holland, Switzerland and Italy. It consisted of twenty leading incandescent lamp factories in Continental Europe. The Association, because of the extreme price competition. established a mutual price regulation, and over-production had proved very unprofitable. The Association therefore fixed better prices, but at the same time, in order to prevent over-production, each company was fixed to a previously agreed quantity, i.e. a fixed quota of lamps. Financially, the results were very satisfactory, but on the other hand any expansion was curtailed.

Meanwhile, in Britain, the lamp industry influenced by its American partners was slow to adopt the new metal filaments and in 1905 attempted to protect themselves from the Continental producers by establishing the *'British Carbon Lamp Association'*. It consisted of British Thompson-Houston (BTH), Cryselco, Ediswan, GEC (Robertson Lamps), Pope's Electric Lamp, Siemens Brothers Dynamo Works and Stern Electric Lamp Company. This Association tried to copy the Continental and American pricing and quota agreements, but the legal position in Britain was more difficult and never had the same success. Sales within this Association were on a contract basis between individual manufacturers and individual distributors, the manufacturers maintaining minimum prices up to the wholesaler retailer but not beyond.

In the meantime, a large variety of carbon filament lamps followed; but the key to success was the manufacturer's ability to develop and produce high quality vacuum, glass to metal seals, glass manufacture, and elimination of contamination/degassing of materials in the lamp. Perhaps most important of all was the material for and construction of the filament. G.E. of America produced an improved version of the carbon filament in 1904 which it used widely in America; it was adopted by the British associate company Thompson-Houston, but it was never a real success in Britain or Continental Europe. In Europe, experiments were carried out on several metals; one notable metal was osmium, which is a rare and difficult metal to work, but it has a melting point of 3,000°C. The lamps proved to be expensive and were very fragile. Some lamps were made with an alloy of osmium and tungsten (in German wolf*ram*), the combination being *Osram*. Whilst the lamps were not a success, the name did survive to become one of the most famous names in the lighting industry.

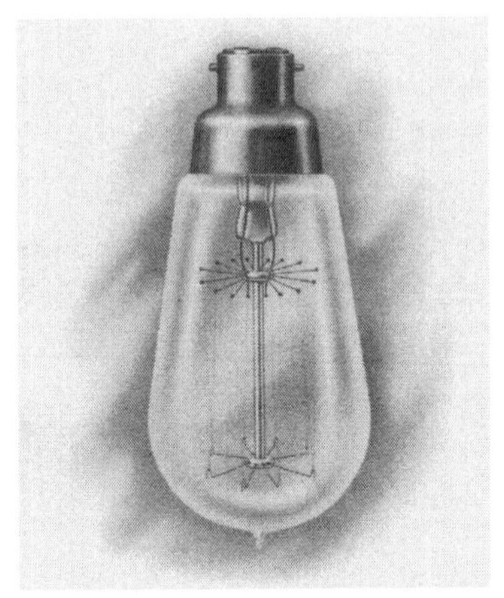

Figure 7: An early Osram lamp with a squirted filament

Perhaps the most significant development, finally destroying the gas mantle as a competitor, was that of Hans Kuzel in 1905 to mix tungsten powder with a binder, extrude it through a very small opening (a die), heating the resulting wire and sinter the powder. This filament became known as the 'squirted tungsten filament', producing twice the light output per watt as the carbon filament. Commercial production of these metal filament lamps started in 1907, but the long fine tungsten wire was also very brittle and research continued to find a way to draw tungsten wire from the solid metal. Probably the first to be successful was the Siemens & Halske Company, who found that an alloy of tungsten and nickel could be drawn into fine wires at room temperatures. This filament was mounted on a system of support wires derived from the tantalum lamp. These lamps were the first to be made with a drawn filament and were called Wotan lamps – from wolfram (German for tungsten) and tantalum. *Wotan* lamps were first sold in 1910.

In America, William Coolidge had been experimenting with a process for making ductile tungsten. Coarse powder was pressed into a bar, which was sintered at a very high temperature by passing an electric current through it. The cross section was reduced by a hot hammering process (swaging), followed by hot drawing through many dies to produce thin filaments of pure tungsten. This new pure tungsten was a new metal, being eventually used in a number of industries but it can withstand much higher temperatures, resulting in a much more intense whiter light in the incandescent lamp. GE of America was the first in production with this pure drawn tungsten filament before the end of 1910. Because of the higher operating temperature of the filament, up to 10 lumens per watt could be achieved. This became the basis of tungsten wire, as we know it today.

The tungsten filament was to bring about the next significant changes in the industry structure by 1907, when the *metal filament lamp* made its appearance. Tungsten, with its higher melting point of 3410°C, higher than any other metal, became the material to be used for the filament. It is extracted from the minerals scheelite (calcium tungstate) or wolframite (iron manganese tungstate). The ore is treated chemically to produce tungsten trioxide and metallic tungsten is obtained by reducing the oxide with hydrogen. Tungsten filament lamps were first produced in Vienna by two scientists Alexander Just and Franz Hanamann, using a sintering process similar to that used for osmium lamps; they licensed the process to a leading Hungarian producer in Budapest. The manufacture of these new tungsten filaments was much more difficult and required more labour; only the larger companies with development laboratories were able to start making their own filaments. The original Continental trade association based on carbon lamps, which had proved to be of great value in the first six years of its existence, gradually lost its importance after 1909. This was because the early metal filaments, which saved approximately 60% of the energy, were being made stronger and cheaper. Naturally, it progressively replaced the much less efficient carbon filament lamp. The Association survived for another four years and finally ceased in 1913.

Already by the start of the British Carbon Lamp Association in 1905, Robertson Lamps was producing more lamps than their major rival, Ediswan. Hugo Hirst bought the British rights to the Nernst lamp and the osmium lamp, neither of which was a lasting success. However, when he heard of the tungsten filaments being developed on the Continent, he was very quick to visit the two scientists Alexander Just and Franz Hanamann in Vienna, and send technicians to visit the factory in Budapest.[25] They had produced the first tungsten filament by heating a carbon filament in an atmosphere of tungsten oxytetrachloride, whereby an exchange takes place between carbon and tungsten. He later went himself to Austria and Hungary to negotiate for what was undoubtedly a major advance in carbon filaments. However, before agreement was reached, Hirst heard of Dr. Auer von Welsbach's (Auergesellschaft) filament development, which he went to investigate; but it was complicated by litigation over German and Austrian patents. Hirst found a clever solution by inviting the warring parties to London, with himself as the mediator, and after some weeks of discussions it was agreed to set up a British Company – Osram Lamp Works in which all three parties had an equal share. A new factory was constructed in Hammersmith next to the Robertson lamp factory. This would produce both the Welsbach lamp using an osmium filament, and the Just/Hanamann lamp using a tungsten filament.

Whilst waiting for production to start, Hirst, to the surprise and shock of competitors, bought one million lamps from Auergesellschaft's Berlin factory (the Welsbach lamp). With some promotion these lamps soon sold, and it was soon clear that the

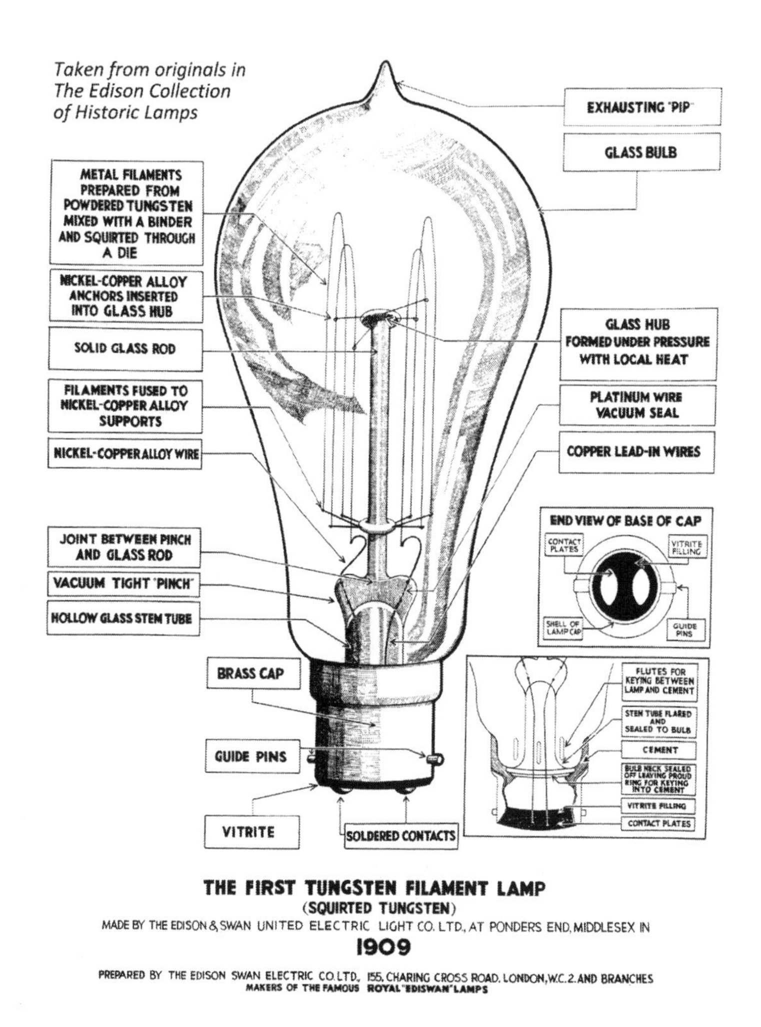

Figure 8: The First Tungsten Filament Lamp (using squirted tungsten)

tungsten metal filament lamps were better products. This eventually led to Welsbach selling his share equally to the other two, making Osram Lamp Works a 50/50 company between Auergesellschaft and GEC. With the success of the tungsten lamp, the support of the German technology and Hirst's sales flair, Osram Lamp Works gained a strong position. Again, Hirst proved himself not only a good marketeer but also a shrewd businessman. The successes of the Osram lamp gave the industry new impetus and encouraged other electrical companies to research and patent a great number of new inventions. In 1916 Hirst took the opportunity to purchase the 50% share of Auergesellschaft in Osram from the Custodian of Enemy Property for £260,000, hence making it a wholly owned subsidiary of GEC.

Product developments and their subsequent patents continued to have a great influence on the industrial scene; GE of America invented the gas filled lamp in 1913 for larger sized lamps. But in 1914 it was discovered that commercial lamps of lower sizes (wattage) could be filled mainly with argon with only a little nitrogen (approximately 10-20%). The gas reduced the heat loss of the filament, and while argon was much more efficient than nitrogen, being denser, it had a lower thermal conductivity. In case of filament failure or impurities in the lamp, an arc could form causing damage to the lamp holder or lamp circuit. The introduction of a little nitrogen prevented this occurrence. In addition, it was also discovered that the rate of cooling of the filament was proportional to the length; hence by coiling the wire, the lamp could draw more current to give more light. Previously the tungsten filament in a vacuum gave a lamp efficiency of approximately 1.25 watts per candlepower (9 lumens per watt) but with the combination of the coiled filament and the gas filling, the lamps became more than twice as efficient, the claim being that a half-watt produced one candlepower. Hence these lamps became known as *'Half-Watt'* lamps; see Figure 9.

Up to this time argon had no commercial value and its manufacture had not been developed; repeated attempts to produce the gas on a commercial scale in Britain had not been successful. In the meantime Philips had discovered a process and was manufacturing argon-filled lamps in large quantities. These lamps were known as 'Half-Watt' lamps, and in spite of several approaches made by the British manufacturers to supply the essential parts of the argon plant, Philips remained reluctant to supply their competitors with this machinery. Eventually a contract, approved by the Board of Trade, was entered into, providing for the supply of an argon plant in consideration, inter alia, of the purchase of a large number of argon-filled lamps supplied by Philips. After unforeseen delays, the plant was received and installed at the GEC lamp works in Hammersmith.

Philips pursued this development during the war and emerged in a strong position, from which the British manufacturers were forced to purchase quantities from Holland. The war had also changed the position of GEC in that it had been able to buy out the Auergesellschaft's share in the Osram Lamp Works from the Public Trustee who had acquired it under the Trading with the Enemy Act in 1916. Also,

Figure 9: Advertisement for the Osram Half-Watt lamp

Siemens Brothers had been sold to a financial syndicate in the City in 1917.

Although the Osram name was registered in the Imperial Patent Office in 1906 as the official trade mark for the Auergesellschaft for its incandescent and arc lamps, it did not become a registered company until 1919. It was then that Osram GmbH Kommanditgesellschaft, registered in Berlin, came into existence by the amalgamation of Allgemeine Elektricitäts - Gesellschaft (AEG), Deutsche Gasgluhlicht - Aktiengesllschaft (Auergesellschaft) and the electric lamp production facilities of Siemens & Halske. This then later became the third great world name in the lamps industry, joining that of GE and Philips.

GE also had a significant influence on the British industry from another direction. In 1928 BTH, a subsidiary of GE, made a successful takeover bid of Metropolitan Vickers (Metrovic) to form Associated Electrical Industries (AEI); this brought the two key lamp brands Mazda and Ediswan under AEI and still very much under the influence of GE.

Huge improvements in processes, in particular the coiling of the filament in 1912, and the coiled coil (double coil) in 1933, led to the lamp we have today, but efficacy is limited to 13 lumens/watt.

English Electric, formed in 1919, mainly involved in heavy electrical engineering, had acquired Dick Kerr, also involved in heavy electrical engineering. However they had a lamp factory in Preston called Britannia Lamps that was very profitable, and in 1927, when English Electric needed capital, they sold 50% of the lamp business to Siemens.

The merging of Electric Lamp Service and Atlas Lamp Works in 1936 was the start of the Thorn Electrical Industries, which, after the start of the war (1940), acquired Vale Lamp Works. Thorn under Atlas started the manufacture of fluorescent lamps in the early 1940s, and in 1949 made a 'know how' agreement with Sylvania of America. E.K.Cole (brand Ekco) acquired Ensign Lamps

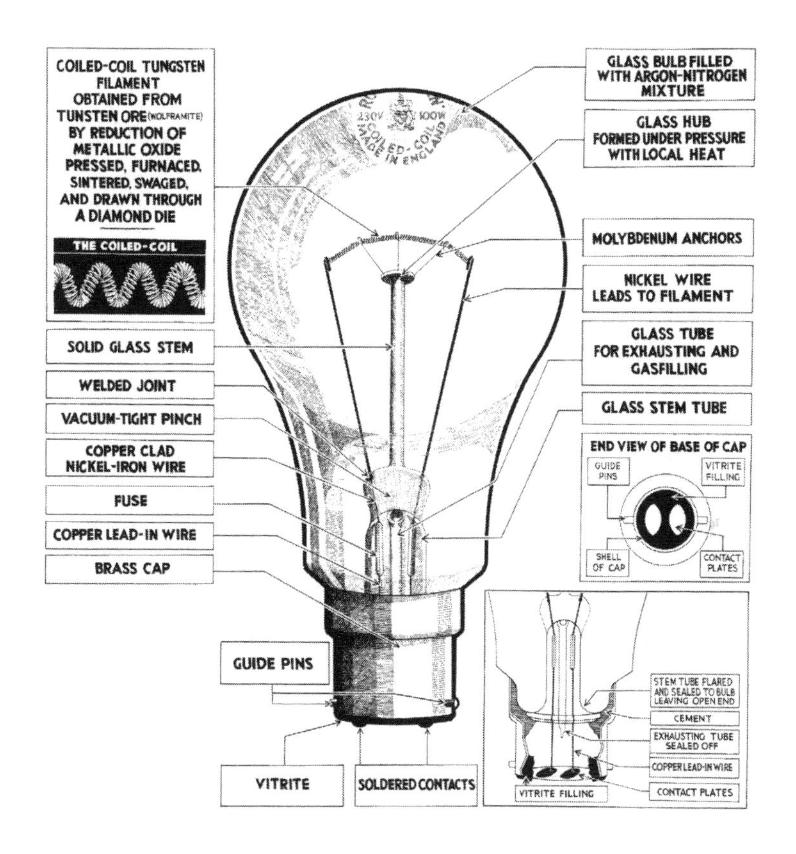

Figure 10: Diagram of a modern tungsten filament lamp

in 1946 and began a close working relationship with Thorn, who later acquired E.K.Cole itself.

[1] The Electric Lamp Industry – Technological Change & Economic Development from 1800-1947, by A.A.Bright
[2] The History of Philips by A.Heerding, vol.2-1986, translated by D.S.Jordan in 1988
[3] The History of Philips by A.Heerding, vol.1-1980, translated by D.S.Jordan in 1985
[4] From the Beginning by Henry Loring in 1905
[5] Sir William Siemens: A Man of Vision
[6] Report from the Standing Committee on Lighting by Electricity, 1879
[7] 'The Times' 12th October 1878
[8] Siemens Brothers by J.D. Scott
[9] Lengthening the Day by Brian Bowers – 1998
[10] The Electrician , 'Incandescent Lamp Manufacture – Flashing' 1887
[11] Electrical Review 'The Edison Filament Case' by J.Swinburne, 6.8.1886
[12] C.H.Stearn memorandum to his staff 26.8.1889 Siemens Archives, no.35-16 Lh 474, Munich
[13] The Electrician 19.7.1889 and 26.7.1889 (report to shareholders)
[14] The Electrician 2.8.1889
[15] Report to the Shareholders of the Anglo-American Brush Light Corporation Ltd., 1887
[16] The Electrician 1.6.1888
[17] The Electrician, 20.7.1888 'Judgement of Mr. Justice Kay'.
[18] Electric Lamp Industry by A. Bright
[19] The Electrician 24.8.1888 and 21.12.1888
[20] The History of Philips by A Heerding, vol. 1-1980, translated by D.S.Jordan in 1985
[21] The Electrician 22.2.1889
[22] The Electrician 20.7.1888
[23] 'Proceedings at the Eighth General Meeting, 13th March 1889, Brush Archives, Loughborough
[24] Minutes of the Board, 1893
[25] Findings and Decisions of a Sub-Committee appointed by the Standing Committee on Trusts (1920)

3
Industry Starts to Consolidate

Not surprisingly, as with so many aspects of the 20th century, the two world wars had a huge influence in shaping the electric lamp industry. The 1914-18 war, so often referred to as the 'Great War', did have a great influence on the lamp industry.

As covered in the previous chapter, the period before 1914 had been very active in product developments of the incandescent lamp. In particular, the metal filament and the gas filled lamp (half-watt lamp) were not only closely protected by patents and licences, but were naturally used as commercial tools to gain competitive advantages and market share. These patents were used as a means of either driving competitors out of the market or controlling them by licence agreements. There is little doubt that the powerful businessmen/financiers behind Edison General Electric Company and the Thompson-Houston Company, which in 1892 became General Electric of America, were the driving forces in the future development of the incandescent lamp market. As the initial patents expired, in 1893-4, many small factories started to produce carbon filament lamps. GE did not worry about these competitors although they formed themselves into the National Electric Lamp Association. It was not until some of them formed the National Electric Lamp Company in 1901, found themselves short of funds and cleverly decided to approach GE for financial support on the basis that healthy competition was good. It was a shrewd move, and Charles Coffin of GE became hooked on the concept, not only providing a substantial amount of capital.

National Electric Lamp Company capital structure consisted of $500,000 ordinary shares and $150,000 preference shares;GE acquired $360,000.[1] It was even more surprising that Coffin agreed to waive the voting rights and a seat on the board. Later the American anti-trust bureau did not regard it as credible that the GE stake in National Electric Lamp Company was made for purely idealistic motives, i.e. to create healthy competition.

Hence in 1910, when the Federal Government brought a case to court, it found that although GE had waived their voting rights to its shares, there was an infringement of the anti-trust laws. To the amazement of all, the learned judge ruled that GE should acquire the entire share capital of National, thereby, in his view ending the illegal practices for which it was accused. To the surprise of onlookers at the time and since, it was through the anti-trust legislation that GE regained almost total dominance of the incandescent lamp industry in America.

Needless to say the Federal Government continued to monitor activities in the American market, and further anti-trust measures were taken in 1911 against GE, Westinghouse, National and thirty other companies, alleging infringement of the Sherman Anti-Trust Act.[2] They alleged that the companies had arranged prices amongst themselves, carved up the market and bought up patents, with the objective of limiting free competition. They were also charged with deluding the American public into believing that GE and National were independent companies competing with each other. Again this had a surprising outcome in that the judge ruled that GE must take over the final 25% of the shareholding of National. So again, the result of an anti-trust case being brought under a law that was intended to prevent monopolies had in fact strengthened the GE position in the lamp market, consisting of 50 million lamps at the start of the 20th century. This totally dominant position of GE in the American market was established, and has continued to remain so throughout the 20th century; and is still the case today.

Not only in America was there an insatiable appetite for incandescent lamps being satisfied by ever-increasing production; it was so also in Europe. Of course the market characteristics of Europe and America have always been very different. Europe is more fragmented, with smaller companies dominating their individual countries with some export activity. The notable exception to this is Philips; their home market was too small for their ambitions and they developed a strong export activity from the early formation of the company. Germany, where most development was being carried out, mainly led the European lamp industry. The main German electrical companies expanded steadily throughout the 1890s, bringing Allgemeine Elektrizitäts-Gesellschaft (AEG) and Siemens & Halske and Auergesellschaft and several other smaller firms to important positions in the period 1899-1900. Unfortunately a depression crept up on them for which they were unprepared, and many of the smaller companies had over-expanded and were forced to liquidate; even the larger ones had to draw on their reserves.

The German electrical manufacturers turned to consolidation as a solution for their problems. Several amalgamations followed, which concentrated most German electrical manufacturers in AEG, the Siemens-Schuckert group, the Felten-Guilleaume-Lahmeyer Werk, A.G., and the Bergmann Elektrizitäts-Werk. In this period of consolidation AEG took over Vereinigte Elektrizitäts Gesellschaft of

Berlin, which had previously been associated with the American Thomson-Houston Company; hence, this acquisition brought AEG into a close relationship with GE resulting in an exchange of patents and limitation of markets. Indeed it also resulted in an agreement with the British Thomson-Houston Company regarding the limitation of exports.[3] While the heavy engineering side of Siemens & Halske amalgamated with Schuckert, the light engineering including lamps remained unaffected. In 1910, AEG took over Felten-Guilleaume-Lahmeyer, creating an even larger combine.

In 1903 there still remained many smaller lamp producers in Germany, and many of them established a close working relationship, to raise lamp quality and maintain prices. It was from this that the organisation grew called Verkaufstelle Vereinigter Glühlampenfabriken Gesellschaft (an association of lamp manufacturers) which became an incandescent lamp cartel with a formal agreement signed in Berlin in 1903; this was under the leadership of AEG and Siemens & Halske. At the time of signing it included eleven lamp producers in Germany, Austria, Hungary, Holland, Switzerland and Italy. It consisted of most of the principal lamp manufacturers in Europe, producing about 30 million carbon filament lamps. The cartel was effective from 1904 and continued more or less until the 1914-18 war. Its principle tasks were to fix lamp prices, establish quotas for the various members and divide the profits. It was only concerned with carbon filament lamps on which there were no basic patents; at that time there were virtually no other types of incandescent lamps. Although Auergesellschaft had introduced the osmium lamp in 1902, AEG introduced the Nernst lamp and Siemens & Halske the tantalum lamp, both in 1905; the market pressure significantly reduced the industry profitability. Besides the new metal filament lamps and price reductions, competition from new firms from outside the cartel and the depressed effect of taxation on lamps were the main contributors in the declining profitability on carbon filament lamps.

By 1910 the German manufacturers (plus those of the cartel outside Germany) produced 26 million carbon filament lamps, 42 million metal filament lamps and 0.25 million Nernst lamps. This was not as great as the American production but considerably greater than the American production of metal filament lamps. By now there were eighteen manufacturers in the cartel covering Germany, Austria, Hungary, Sweden, Holland, Italy and Switzerland. A few French manufacturers joined but withdrew after a short period and the British producers never joined at all. In 1911, the three main German producers AEG, Siemens & Halske and Auergesellschaft formed the Drahtkonzern (Filament Trust), through which they pooled their patent rights. Under German patent law and the interpretation of the German courts it was much harder to obtain a complete patent monopoly than it was in America or Britain.[4] In these circumstances, it was natural for the owners of German patents covering all the ways of making tungsten filaments to pool their patents and obtain basic protection. This remained the central European position up to the start of the 1914-18 war.

Unfortunately, from 1885 to 1905 British manufacturers fell far behind the Germans and the other major European producers. Initially, because of the patent protection position being adopted by the Edison & Swan United Electric Co., much of the industry innovation was driven abroad, but more importantly it imposed restrictions on electricity expansion. While the great legal obstacle to electricity expansion was removed in 1888, what remained for nearly 20 years was apathy, leading to a lack of specialisation, little investment, and little or no contribution to the development of the metal filament lamps. Indeed, there are no records of any lamp research laboratories in Britain during this period. Instead, the British industry relied upon American and German producers who had subsidiaries or associated companies in Britain, and who held the British patent rights of their parents. They imported and marketed lamps made by their parents or had manufacturing operations under their parent companies. From 1901, GE had a controlling interest in the British Thomson-Houston Company. Westinghouse formed three companies in all, the first being in 1889: the Westinghouse Electric Company Ltd to handle patent rights. The second was in 1899: the Westinghouse Electric & Manufacturing Co. Ltd. and the third was in 1906: the Westinghouse Metal-Filament Lamp Co. to work with the Austrian Westinghouse company in the marketing of tungsten filament lamps. The German Siemens & Halske exerted its influence through its associated company Siemens Brothers Co. Ltd.

GEC remained independent, and indeed became the market leader after the turn of the century, being the most aggressive competitor and the first to secure the rights from AEG to the Nernst lamp and the Austrian developed osmium and tungsten lamps. Having acquired the ownership of the basic British tungsten lamp patent in 1904, the work of Alexander Just and Franz Hanaman was probably the key factor in consolidating its position as the market leader in Britain for decades to come. As indicated in the previous chapter, the production of carbon lamps was through the Robertson Lamp Company and for the tungsten filament lamps, the Osram Lamp Company (this being jointly owned with Auergesellschaft of Germany). Two of the early pioneer lamp companies were still in business, i.e. Edison & Swan Electric Light Co. Ltd. and the Sunbeam Lamp Co. Ltd. plus a number of smaller newer companies of varying size and importance.

Soon domestic competitors and importers of foreign manufacturers started to introduce their own tungsten lamps into Britain on a large scale. This sparked off a new round of lawsuits by GEC and Osram Lamp Works to test their patents, probably the first most important one being instituted in 1910 against G. M. Boddy & Co., an importer and distributor of lamps made in Holland by Philips. Before the lawsuit came to court, an agreement was reached by which Philips and Boddy took licences under the GEC patents and agreed to pay royalties on all lamps imported to Britain and to limit the total quantities imported.[5] Several other infringement proceedings by GEC were similarly successful. Prices and discounts were also to follow those set by GEC. Unlike the Ediswan Company some twenty years earlier, GEC required other manufacturers to take patents licences and pay royalties rather than trying to force them to withdraw from the market.

Two other British companies held important tungsten filament licences; the British Thomson Houston Company (BTH) owned several patents based on the work of GE of America and the Siemens Brothers owned patents based on the work of Siemens & Halske in Germany. With the original Carbon Lamp Association becoming less effective, in 1912 these three companies, GEC, BTH and Siemens Brothers formed a new Tungsten Lamp Association. In this Association they pooled their patents and granted licences to other companies including Philips, Ediswan and British Westinghouse. These companies agreed to maintain selling prices and those which were only licensees were required to pay royalties and to remain within agreed production quotas. This Association established an indirect affiliation with the German Drahtkonzern through the British subsidiaries of the German companies, which were members.

Hence it was through the very astute moves of Hugo Hirst of GEC in the early years of the 20th century that re-established Britain alongside Germany as leading European industrialised lamp countries. The French industry had declined; their market for carbon filament lamps had been quite open, creating keen competition and forcing prices to very low levels. The leading American, German and British lamp producers had subsidiaries or affiliates in France. There had been fewer consolidations at that time in the French market than in the other main markets. While the Austrian industry had established considerable importance technologically, it was quite small. The producers at this time in Belgium, Hungary, Italy, Sweden and Switzerland were not of great importance to the world industry except through their participation in the international cartel (this is covered in detail in the later chapter on trade associations). The possible exception was Philips of Holland, though having a small home market it was beginning to establish strong export activities and it was also quick to establish tungsten filament lamp production.

The Trend in Improved Performance of Incandescent Lamps 1881-1910[6]

YEAR	TYPE OF FILAMENT	INITIAL EFFICIENCY LUMENS PER WATT	APPROX. USEFUL LIFE (HOURS)	
1881	Carbonised Bamboo	1.7	600	
1884	"Flashed" squirted cellulose	3.4	400	
1888	Asphalt-surfaced carbonised bamboo	3.0	600	
1897	Nernst (refractory oxides)	5.0	300	on DC
			800	on AC
1898	Osmium	5.5	1000	
1902	Tantalum	5.0	250	on AC
			700	on DC
1904	GEM (metalised carbon)	4.0	600	
1904	Non-ductile tungsten	7.9	800	
1910	Ductile tungsten	10.0	1000	

NB: Efficiencies apply to the most commonly used lamps, 16 candlepower for the carbon lamps and 50/60 watt for the GEM and later metal filament lamps.

As with many industries, the early technical developments were key factors in its formation and the early lamp industry was shaped not only by astute businessmen but also in the race for patents and the ability to defend them and use them for licences, hence royalties. With the products being commodities in today's terminology, the manufacturers were very competitive from the early days of the industry. Already from the early years of the twentieth century, the larger more innovative companies were in a position to sell components to competitors to enable them to exercise more influence. In these early years the production of glass, as well as the filaments, was very important. Inevitably these situations did bring these industrial leaders of the various producers into contact, leading to a series of agreements to stabilise the markets; this is covered in Chapter 5, Trade Associations.

Indeed, it could also be said that mergers had a strong influence in Britain through the Edison & Swan United Electric Company and by the actions they took in the 1880s and 1890s to protect their patents and drive their competitors out of business. Furthermore, it had a similar influence in Europe as Edison/GE began to make their bid to dominate as they had they had in America.

Of course, the market characteristics of Europe and America have always been very different, Europe being more fragmented and with smaller national companies dominating their individual countries, with some export activity. In Britain GE tried, through the British Thompson-Houston Company, to exert its influence. Indeed in 1911, BTH acquired from GE all the patents for drawn wire tungsten filaments and the Mazda trademark. By 1913 the BTH profits were at an all-time high, almost certainly due to the growth in the lamp business. GE established a good relationship with GEC and Siemens Brothers, but they kept steadfastly out of the American influence.

Throughout Europe, GE had established companies in the major countries, i.e.

Britain:
British Thomson-Houston Company.

Italy:
Thompson-Houston Società Italiana di Electtricità
Societa Edison-Clerici

Spain:
Thompson-Houston Ibérica
Societa Edison-Clerici

France:
Compagnie Générale des Lampes Incandescentes (the merged Edison and Swan companies)
Thompson-Houston Compagnie Française

Germany:
Deutsche Edison Gesellschaft (AEG)

Japan:
Tokyo Electric

Thus by 1914 GE had a commanding foothold through their shareholdings in the world lamp business, putting them in a powerful and influential position with many restrictive agreements throughout the international industry. Perhaps the main exception was Philips, whose market influence and financial strength had grown significantly. The effect of GE's influence in Europe was very significant, having established an exchange of patents and know-how with its European partners. It divided the European lamp industry into three geographical blocks:

The French, Spanish and Portuguese markets were dominated by the French Thomson-Houston company.

The rest of Europe (except Britain) came under the influence of the Patentgemeinschaft (patent pool) that was dominated by the three main German companies AEG, Siemens & Halske and Auer. Many small and medium sized lamp-makers throughout Europe, with no access to the GE licences and no technical support, had to acquire their new production technology; this was both costly and time consuming. Even then they faced possible conflict with the patent holder. Hence most of these companies fell under the influence of this powerful grouping.[7]

In Britain the British Tungsten Lamp Association was formed in 1912, being an alliance between GEC, British Thomson-Houston (BTH), Siemens Brothers and Ediswan. All had been members of the British Carbon Lamp Association, but now with the advent of the metal filament had to improve their manufacturing techniques whilst avoiding litigation among themselves. Hence they entered an agreement to pool their metal filament patents. Their declared objectives were to promote and protect the interests of manufacturers and dealers of electric lamps in the UK. Also to carry out research for the improvement of electric lamps, agreements were entered into with members, wholesalers and retailers relating to the manufacture, supply and sale of electric lamps.[8]

In the period up to the 1914-18 war, the American industry became dominated by GE by various take-overs and amalgamations, but in Europe no one company or any one nation secured all the most important domestic patent rights covering lamps; consequently no one company could dominate the industry. Control was exercised through the formation of patent pools chiefly in Germany (covering much of central Europe) and in Britain. It could be said that competition was very keen with enormous pressure on costs, which had harmful effect on quality and most likely reduced the emphasis on development, with the consequential effects on profits, until the formation of these patent pools.

Production had expanded at a tremendous rate both in volume

and value, with the manufacture of lamps emerging as a mass production industry. Production methods and lamp design were considerably improved and a large variety of types grew for many special applications. The new filament materials, particularly tungsten, brought much improved lamp efficiency guaranteeing the future of the industry. Gas lighting was unable to advance beyond the high pressure Welsbach mantle and began to fade in relative importance. Arc lighting kept up for some time, but it too began to be replaced more and more by the simpler and improved incandescent lamp.

Naturally, the outbreak of the 1914-18 War placed considerable strain on many British electrical companies.

During the war, the Government took over the shareholdings held by German companies in essential industries; this applied in particular to Siemens Brothers and Osram. In the case of Siemens Brothers relationships with Siemens & Halske were effectively severed on 4th August 1914. There had been no precedents for handling situations of this nature on the outbreak of war; however, the Government, on 5th August 1914, issued a proclamation prohibiting trade with the enemy. While one of the directors, Sir Walter Lawrence, applied to the Foreign Office for guidance, the situation soon became very clear with the resignation and subsequent internment of the managing director of Siemens Brothers Dynamo Works, Karl von Köttgen. Though he was released some weeks later and there were obviously difficult times with many conflicts of loyalties, Sir Walter Lawrence resigned from the Board and the legal position remained unclear. In January 1916 the Government passed the Trading with the Enemy Act, which effectively meant that the State appropriated enemy-owned property. In the case of both Siemens and Osram this effectively meant that the Government took over the shareholdings belonging to their German associates. This was vested in the Public Trustee, who, in 1917, sold that value of share capital previously owned by Siemens & Halske to C. B. Crisp & Company; this position remained to the end of the War.[9] Hence a new board was created involving business men, a financier from C. B. Crisp & Co., and a solicitor; only one of the original board remained, this being Von Chauvin, who had successfully applied for naturalisation at the beginning of the war. This created a board without specialist knowledge in the electrical or lamp making businesses, but they had experience in business, finance and public life and the board was much more outward looking than previously. Following the arrangements made with Crisp and the Public Trustee, the company had a huge debenture liability and immediately after the war it had to raise new capital to reduce this debt. New capital was raised and the Board not only reorganised the capital structure of the Company, but also trimmed the organisation to meet the tough market conditions following the war.

The recommendation of a post-war Government Committee included the following comment:

> While America, Germany and other countries were eagerly seizing the benefits of electricity, our authorities were busy with the erection of obstacles to its development. Thus for years this country stood still, leaving the field open for foreign manufacturers to gain, both in their own and other markets, a hold which they have never lost. Abroad there was strong encouragement, in Great Britain irksome restrictions, and the

> evil effects on the British electrical industry justly entitle it to now special remedial measures, which can be secured only by legislation. Such legislation should be passed without delay.[10]

So indeed the scene was set in this post-war period for a more sympathetic environment for the whole electrical industry, including the electric lamp industry. But even as late as 1925 there was still unease in government circles, concerning Britain's backwardness in electrical development. The then Prime Minister, Mr. Baldwin said '... in this country we have lagged behind' and he referred to a committee under Lord Weir, which had been appointed to consider what could be done. In January 1926 he outlined the plans for the Electricity (Supply) Bill, which was to rationalise and increase the output of electrical power. Electrically, the face of Britain was to be transformed. The then Chancellor of the Exchequer, Mr Winston Churchill, said ' the creation in this country of a system of electrical supply for all purposes which will be unsurpassed in the whole world is not a luxury, but a prime necessity for the life and well being of our whole country ... we are going to create it without any further delay.'[11] The Bill was introduced in March 1926 by the Minister of Transport, Colonel Ashley, who in the second reading of the Bill remarked 'In truth it may be said that only about one third of Great Britain is reasonably supplied with electricity.'[12] Finally, the blocks were in place for the real expansion of the electrical industry as a whole.

In 1916 Dudley Docker, a wealthy financier and businessman, had founded several public companies including Metropolitan Carriage Wagon and Birmingham Small Arms; perhaps more significantly he founded the Federation of British Industries. He had ambitious plans to create a giant British electrical company to rival those in

America and continental Europe. He was a passionate supporter of British industry, a great advocate of mergers and he attacked those people who feared that these large combines were to the detriment of the public good. Docker was not only convinced but often spoke in public on the virtues of large industrial combines to provide stable wages; and he became a pioneer of issuing shares to workers and in establishing works councils.

Many of the large electrical companies throughout the world had been established on the back of the electric lamp industry which, in itself, became a key driver for the spread of electrical supply, particularly to homes. Dudley Docker had seen the significance of this. While other electrical products and electrical supply equipment had become much more dominant, the lamp businesses maintained an important and profitable role due to patents, licences and distribution strength to the electrical trade. Indeed, Siemens did own several important patents and licences for lamps as well as significant cable production facilities and it was one of the leading members of the lamp ring of which both GEC and English Electric were both keen to acquire a larger share. When, in 1916, British Westinghouse had been refused admission to the Federation of British Industries, because it was American controlled, Docker must have played a part in this rejection. The Board began to consider whether the American ownership was more of a hindrance and a holding company was formed to purchase the American shares. The man the board turned to for help to achieve this was Dudley Docker; thus later, British Westinghouse became independent from American control. It was after this in 1917 that he persuaded the board of British Westinghouse to set up a committee to study the possibility of collaboration with Vickers; however, it recommended that there were no advantages in such collaboration.

Whilst these companies had diversified into other electrical products, any restructuring was bound to influence the formation of the future lamp industry. Docker's plans were to establish British Westinghouse at the centre of an industrial combine; he let it be known that he wished to purchase Siemens Brothers (including Siemens Dynamo Works) from the Custodian of Enemy Properties. This never came about, but the British financial group headed by Birch Crisp, now effectively the owner of Siemens Brothers, sold the Siemens Dynamo Works to the newly formed English Electric Company in 1919. Along with this, they signed a twenty-five year agreement by which 'the classes of business carried on by Siemens and English Electric' were 'exclusively reserved to each of the parties'.[13] This effectively locked each company into being the preferred supplier to the other.

Docker then turned his attention to combining British Westinghouse with GEC, and although he and GEC's Chairman Hugo Hirst had agreed in principle, the GEC board refused to support it; hence the plan was abandoned. Docker eventually sold Metropolitan Carriage Company (including his controlling stake in British Westinghouse) to Vickers in 1918, thereby creating Metropolitan Vickers, which became abbreviated to Metrovick, and the name 'British Westinghouse' disappeared. They did eventually purchase another ex-German owned company from the Custodian of Enemy Properties, the much smaller Brimsdown Lamp Works.[14] Along with the Westinghouse lamp activity, this established Metrovick as a player in the lamp business.

Having had a difficult, indeed chequered history, British Westinghouse was now under a new umbrella and being mainly managed by the Vickers management. Metrovick was put on a sound financial footing and established itself as one of the four large companies dominating the British electrical industry, the others being GEC, English Electric and BTH.

English Electric had now become an effective force in the electrical industry, being formed out of the combination of Dick Kerr, Siemens Dynamo Works and Coventry Ordinance Works. Throughout these negotiations the directors had established an understanding that English Electric would handle the heavy electrical industry, and Siemens Brothers the light electrical industry. However, Dick Kerr, primarily involved with heavy electrical engineering, had already acquired Britannia Lamps, with its Preston factory, through an earlier take-over. In this way he brought English Electric into the lamp business and created an overlap. In order to resolve any possible future problems in their electric lamp businesses, it was decided to create a jointly owned company, Siemens & English Electric Lamps.

Although formal links between Siemens Brothers and Siemens & Halske had been broken during the war, a new general manager, Dr. Henry Wright, was appointed in 1924; he turned out to be an extremely strong personality, firmly stamping his views and strategies on the Company. Following his technical and scientific education in Germany, he went to work for Siemens & Halske in Vienna in 1902. One of his first actions was to renew the links with the former German parent. Agreements were made for a complete exchange of technical information and cross licensing of telephone patents throughout the world. This eventually led to Siemens & Halske acquiring 15% stake in Siemens Brothers (the new articles of association drawn up after the 1914-18 war prohibited any foreign shareholding larger than 25%). It was not surprising to see that in 1927 Siemens Brothers bought the lamp business from English Electric, who badly needed cash, and ran it themselves. This lamp business was to be highly profitable to Siemens Brothers and was to eventually generate half their overall profits in the 1930s.

During this period, the lamp business of Siemens Brothers maintained a high degree of independence; in 1923 they moved their lamp works from Dalston in London to Preston in Lancashire. Here they continued their small development activity initiated when links with their German parent company were severed in 1914. Although small, the development activity at Preston proved to be highly successful, particularly under the direction of Dr. Wright and the renewed links with the previous German parent. It emerged into a leading position alongside the large national and, indeed, international companies.

While it was seen as logical, when, in 1927, English Electric urgently needed cash, they sold their 50% of the very profitable lamps business to Siemens. The mid-1920s industrial depression affected all industries, although the lamp market had probably suffered less than most other industries. For English Electric and the then managing director, George Nelson, it was a serious mistake as during the 1920s and 1930s it was the lamp businesses which proved to be the highly profitable, the most cash generative part of the large electrical companies. Indeed, according to his son, George Nelson had tried on several occasions to purchase a lamp company in the 1930s. His son also recalled the occasion when George Nelson had his negotiations to acquire a small lamp company frustrated by a last minute better bid from an unknown

Viennese immigrant whose name was Jules Thorn.[15] Neither he (George Nelson) nor English Electric ever managed to get back into the lamp business.

British Thomson-Houston, being owned by GE, had been progressing well during and immediately after the 1914-18 war, had seen the demise of British Westinghouse and were determined not to follow the same route. They expanded rapidly in the early 1920s into a variety of new electrical products, which inevitably meant several new factories for which new capital was needed. This was mainly supplied by GE but they also raised some in the City of London. Despite all this, profits continued to disappoint. The heavy electrical engineering side and the new domestic appliances and radio activities were making little or no profit; indeed, in 1927 the prosperous lamp activity made more profit than the rest of the company, indicating that in reality the other activities were by then loss-making.

It was therefore not surprising that the concept of amalgamation of some or indeed all of the large electrical companies would be again discussed. But this time it was not Docker but Gerard Swope the president of GE who was involved, though the man who initiated the discussions in 1926 was Anson Burchard, head of International General Electric (IGE). From 1921 onwards, Gerard Swope made many trips to London and he was in regular direct contact with Hugo Hirst of GEC, particularly to persuade him to merge with BTH. The Midland Bank, GEC's bankers, and two of their directors Reginald McKenna, the Chairman (former Home Secretary and Chancellor of the Exchequer) and Dudley Docker, were both exponents of amalgamations to create powerful, strong and innovative companies. Gerard Swope lost no time in trying to persuade them of the benefits of his proposed merger(s). Because of the very sensitive nature of such a concept, all discussions were carried out very subversively and only a select number of people knew of Swope's plans. Although there was excess capacity in some areas of the heavy electrical side of the business, Hirst was not willing to consider a merger. Indeed, the first mention of these discussions in the GEC board minutes was in May 1922 in which the Chairman, Hugo Hirst, reported the offer from Swope and reasons why he declined; the board gave their unanimous support.

Swope certainly did not give up and he continued to plan and scheme to achieve his ultimate goal of the merger of BTH, Metropolitan Vickers, English Electric and GEC. He continued to strengthen the GE shareholding in BTH by purchasing any available shares. He purchased two other small electrical companies: Ferguson Pailin and, in particular, Ediswan, which had by this time become a small lamp producer, having steadily declined after the loss of its patents. Continuing his covert planning, in 1927 he secretly bought a controlling interest (78%) in Metrovick from Vickers and registered it in the name of Dudley Docker. When the proposed merger between BTH and Metrovick became public, everybody, including the Metrovick board, thought this was part of the original 'Docker plan' for a great British electrical company; but Dudley Docker had been a well- paid 'stalking-horse'. This covert operation did not come to light until many years later.[16] In July 1928, when Sir Philip Nash, Chairman of Metrovick, reported to the board that negotiations had started with BTH 'with a view to securing co-operation in engineering, manufacturing and selling', there was no hint in the minutes that Nash or any other member of the Metrovick board realised that GE not only controlled BTH but was ultimately controlling their company.

By the end of 1928, Swope had accepted that only two of the four companies, BTH and Metrovick, would merge and adopt the name of Associated Electrical Industries (AEI). In the formal offer document it was stated that GE would have a minority interest; we now know this was clearly untrue and that GE indirectly controlled more that 50% of the AEI shareholding. This still left Swope without a monopoly in Britain, but with a very powerful voice, particularly in the cartel agreements; these were now very important in lamp industry circles to maintain profitability, particularly in America. These agreements are covered in a later chapter. Being a determined, scheming and a very clever man, he was still not about to give up.

It was at this time in 1928 that Swope again applied his mind to GEC; he had also been building up, through various nominees, a large shareholding in GEC, which he clearly regarded as the jewel in the crown of the UK electrical industry. It was known at that time that there had been a lot of buying of their shares on the American market through various nominees. By now, the board had become suspicious and begun to worry that the control of the company could fall into American hands. Hirst agreed with the board to call a special shareholders' meeting in September 1929, to ban foreigners from holding votes. He told shareholders 'whilst we welcome shareholders from whatever nationality it is not merely a financial and industrial undertaking, but it has also a national aspect. For that reason we are desirous that the control should at all times be in the hands of British subjects.' The resolution was carried and voting rights for foreigners (non-British subjects) removed, including those of foreign corporations or corporations under foreign control. Indeed, this gave the directors the right to suspend the voting rights of the shareholders concerned until they could prove they were not subject to foreign control.

However, in the spring of 1929 Swope asked Owen Young, the Chairman of GE, to negotiate with Hirst. These negotiations were again kept a closely guarded secret. The proposals had impressed Hirst and he brought in Max Railing, his managing director, after which it was thought that the plan was acceptable but would need a lot of details to be worked out. A detailed plan was worked out involving a merger between AEI and GEC with Hirst as Chairman. The subsequent internal discussions within the GEC board proved too difficult, the following being an extract from a letter from the GE representative in London to Swope:

> *My observation of the situation and contacts with Sir Hugo Hirst, have left me with the definite impression that while he personally would be willing to see a merger brought about with the Associated Company (AEI), his associates and particularly his immediate colleague Mr. Railing are very much against it. They did everything they could to prevent Sir Hugo from going forward with the plans outlined in my last memorandum of which you have a copy and privately I understand that at a board meeting the GEC which took place on Tuesday, they voted against Sir Hugo's proposal of a merger.*

Swope made other unsuccessful attempts to achieve his grand plan for the British electrical industry. By the mid 1930s, he was under pressure from the Governor of the Bank of England to reduce the GE shareholding in AEI to about 40%. He also sold a substantial number of GEC shares in 1935.

The hand of Gerard Swope and hence GE played a very important role in the early consolidation of the British electrical industry and hence the formation of the lamp company consolidations. Because of his covert ways of working, little was ever known at the time of the influence Swope had, neither by the public nor the directors of the companies involved. His role in the General Patent and Business Development Agreement (more commonly known as the Phoenix Agreement) is outlined in the later chapter on trade associations. There is also little doubt of the important role he played in the development of GE itself, in creating the great company it is today, not only by protecting its home base but also in the expansion itself. With the benefit of history, it is worth noting that following the great industrial take-over battles forty years later, in 1966-1968, Arnold Weinstock achieved almost all of what Gerard Swope originally wanted.

By the end of the 1920s therefore the lamp industry had consolidated into the hands of GEC, AEI, to a lesser extent Siemens Brothers, and a number of smaller companies.

Furthermore, one of the first post-war agreements was with Philips in 1919, arising out of patent litigation won by GE after Philips had started to advance into the United States market just before and during the war. Philips had been constrained by the quota agreements within the Patentgemeinschaft (patent pool), in which they were restricted to selling only about half their capacity within Europe and Philips had decided to use a significant part of the surplus capacity in the United States. In 1920 the outcome of this agreement was for GE to purchase 18.7% of Philips capital.[17] As so often in company strategy, it can be seen that it was driven by one or two men, as demonstrated by the following extract from a letter by J. M. Woodward (European Director for IGE) in 1924 to A. W. Burchard (President of IGE):

> *I desire to effect an arrangement which will be to the absorbing interest of the Philips Company throughout much of the remaining 'useful life' of Anton Philips as possible. He and practically he alone constitute a danger to our American profits. Should our patent protection at any time become weak in all or part of America, he will be our greatest menace, the least vulnerable and most resourceful of all our competitors. He is insanely anxious to get into England, and for my part, if I could arrange it on a sound basis and without losing my personal friends locally; I would get him there in short order. England is the greatest bait, carefully handled, which can be held out to him, and if we exercise the ordinary principles involved in poker, I am satisfied that we can make arrangements which will be profitable and satisfactory for a long time to come.*

Philips had already tried to establish a presence in Britain by acquiring 10% of the share capital of the Edison & Swan Electric Company in 1919,[18] and it was agreed that both Anton and Gerard Philips should join the board. Part of the agreement was for the creation of a separate works for the incandescent lamp manufacturer. Immediately Philips' engineering workshops began to make and ship the necessary lamp making machines. Unfortunately, the installation, operation, quality and working practices fell far below Philips' management expectation, causing friction between the companies. Friction was exacerbated when the Edison & Swan Company fell seriously into arrears in paying for this equipment but insisted on paying a dividend, which the Company's finances could not afford. Anton Philips lost all confidence in the joint venture and had to finally resort to legal proceedings to recover the debt. The company continued to struggle until eventually, in 1925, BTH purchased the whole of Ediswan, including the Philips shareholding.

Anton Philips was not discouraged; in 1921 he opened discussions with Von Chauvin the managing director of Siemens Brothers, but agreement could not achieved because of the complications with the 50/50 shareholding with English Electric.

In the meantime, BTH had been bringing a number of patent actions against customers of Pope and Volt (from Holland), and two lamp manufacturers, Corona Lamp Works and Cryselco (which was a subsidiary of the wire manufacturer, Duram). Quite separately, Corona and Cryselco approached Philips for both financial and technical assistance in their battle with BTH. This involved the 'Half-watt' gas-filled lamp patent, and it was clear to Philips that if BTH won their 'Half-watt' patent against Corona and the drawn tungsten filament patent (known as the Coolidge patent) against Duram/Cryselco, the British Ring would be a monopoly for the lifetime of the patents. It was very clear that this outcome was not in Philips' interest and they agreed to share Duram's costs arising from their current lawsuits and to help Corona defend its case before the House of Lords.[19] Anton Philips struck up a good relationship with the Duram management; even when they decided to drop the case against BTH, due to a boycott of their products, and negotiate an agreement with the Ring, the trust with Philips was maintained. In the following year, 1923, an agreement was signed and Philips took over the technical management of Cryselco. This relationship continued for the next 65 years, until a management buyout forced a decision to withdraw from the lamp business.

Anton Philips, like Gerard Swope, had designs on the British market. This can be seen from this text from a letter to his brother Gerard Philips following a visit from a director of the National Providence Bank to settle the outstanding debt from Ediswan; Gerard tried to persuade Anton to change his mind on the future relationship with Ediswan:

> *If we could do a deal entirely to our own satisfaction, i.e. have a separate company, it would be quite possible to bring in the lamp-making businesses of Siemens Bros. and Vickers at a later stage. We might then be making half as many lamps as the British cartel before very long.*[20]

Having had negotiations with Metropolitan Vickers and Siemens Brothers in 1922, and also with Ediswan, Anton Philips had great faith in technical and organisational superiority but continually faced the challenge of being a foreigner trying to take over traditional British companies.

Continuing through all this was the fierce battle against the Half-Watt patent and the drawn filament patent. At the beginning of 1924 the number of combined writs being served on customers of Pope and Volt was sixty-eight. These costly lawsuits eventually concentrated on one specific action against Charlesworth, Peebles & Co., who were Pope's agents in Scotland. Both the High Court and the Court of Appeal had found in favour of BTH. It was at this point that both BTH and Philips tried to find a compromise, but BTH proposed that Philips should pay 600,000 guilders compensation and were not prepared to negotiate on this figure. Philips calculated that if they continued the fight to the House of Lords the costs would rise to 800,000 guilders; they decided to continue the fight. To the surprise of all, in February 1925 the House of Lords unanimously ruled in favour of Charlesworth,

Peebles & Co. and against BTH. This proved to be an interesting interpretation of patent law as the Law Lords cited a patent case of 1913 in which BTH had similarly claimed the rights in the invention of the drawn tungsten filament. In that case BTH's claim had been based on a hypothetical theory-based patent dating from 1906. The Lords considered this sufficient reason to regard the 1906 patent, imperfect as it was, as being in anticipation of the Coolidge patent of 1909.

Therefore, the Coolidge patent did not fulfil the essential condition of novelty, although it was universally accepted to be a genuine and valuable invention. The Lords found themselves reluctantly compelled to declare the Coolidge patent void.[21] This logical interpretation of the law had created a strange situation which caused the learned judge Lord Dunedin to complain: 'I cannot help feeling that in these proceedings a real inventor has lost the fruit of his invention.'[22]

Being the shrewd man he was, Anton Philips had been negotiating with the Ring before the House of Lords ruling; and by the end of 1924 he had arrived at an agreement to supply, on free sale, 3.2 million lamps per year. There is little doubt of the hidden hand of J. M. Woodward of IGE, who had urged BTH to make a concession to Philips in the hope of getting Anton Philips to sign an international cartel agreement which would restrain the fierce competition in the incandescent lamp business. Woodward regarded Anton Philips as the most aggressive of the international competitors and hence the greatest threat to GE's dominance of the lamp industry. He was already planning the cartel arrangements that were later to become the Phoenix Agreement.

Philips' persistence eventually began to pay off and a bridgehead

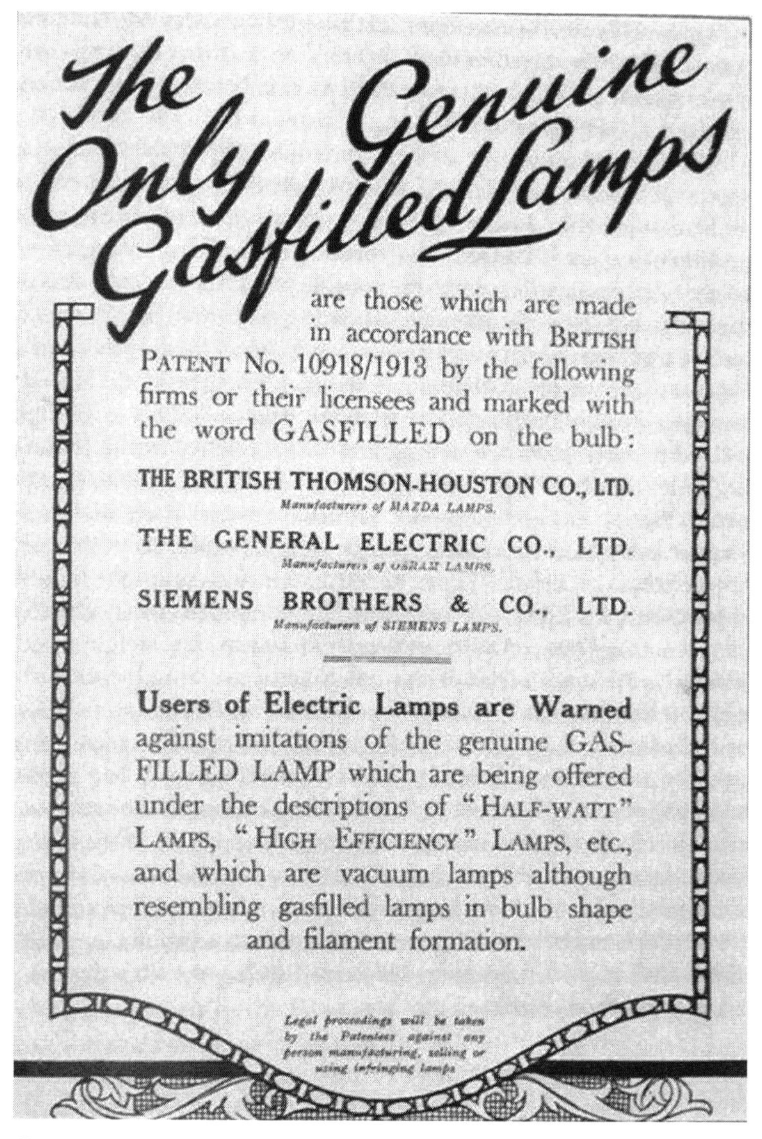

Figure 11: Notice in the Electrical Times on 12th October 1922

had been established in Britain. This gave them a sales foothold into most of the developed world, with the notable exception of United States and Canada, which GE had most fiercely protected. The only other exception was Japan, and Philips deliberately chose not to enter. The foothold in Britain was seen as the opportunity to make inroads into the British Colonies, which would be much easier from Britain than it would have been as an outsider.

Naturally, the British manufacturers were concerned to see the advance of Philips in their market and in 1927 ELMA, the British Ring, brought and won an action under the Merchandise and Marks Act. The complaint concerned the importation by Philips of lamps without an indication of the country of origin, as these should have been marked with 'Made in Holland'. Nevertheless, Philips made steady progress in the British market and some six years later, when ELMA (Electric Lamp Manufacturers Association), the British Ring, was restructured, Philips was to become a member.

In this post-war period, an amalgamation of particular interest was happening in Germany. The three key German companies survived the war but were in a very weak position, with many of their export markets taken or confiscated. Hence, with the various national markets in some state of flux and all the major European companies as well as GE battling to keep or even gain share, inevitably competition was very fierce, and prices became very competitive, bringing many companies into losses. Representatives of AEG and Siemens & Halske had met only months after the end of the war in 1918 to exchange views. A key consideration was a closer co-operation, perhaps even the formation of a sales syndicate; however, their previous experience with the carbon filament cartel, the VVG, had shown that even with a well organised sales syndicate, it was the outsiders, or those who threatened to become outsiders, who benefited the most. After extensive discussions they decided to create a long-lasting solution by merging their incandescent lamp factories.

This was not an easy path for any of the companies. AEG were strong advocates of a structured industrial organisation, created by Walter Rathenau (1867-1922) and carried on by his son Emil, who took over in 1915. The plan had the most opposition from Siemens & Halske, whose lamp division was headed by Otto Feuerlein, who anticipated many technical and business complications from such a merger. He pressed hard to return to the close alliance of the Patentgemeinschaft involving the three main Berlin companies, AEG, Siemens & Halske and Auer (Auergesellschaft).

AEG, frustrated by the protracted discussions, entered into bilateral talks with Auer; in September 1919 they came to an agreement based on the original solution of merging of factories. Siemens & Halske were invited to join these discussions and it soon became clear that the strategic and economic advantages were too attractive, and the Board of Siemens & Halske overruled Feuerlein's objections. Agreements quickly followed, and in November 1919 all three parties set up a joint company, in reality a huge manufacturing operation for incandescent lamps. There were a number of risks in such an operation: possible nationalisation, or the government insisting on worker control. It was accepted that if the electrical industry were at any time to be passed to worker control, then the lamp sector would be among the first to be affected.

In February 1920, the new organisation was formally established under the name of Osram GmbH Kommanditgesellschaft ,being a limited partnership with a capital of 30 million marks. The Osram name had been originally registered at the Imperial Patent Office in 1906 as the official trade mark of Auergesellschaft for incandescent and arc lamps and was registered as a company in Berlin in 1919. In this newly formed organisation, AEG and Siemens & Halske each had 40% of the shares, the remaining 20% being held by Leopold Koppel, the proprietor of Auergesellschaft. Division of shares in proportion to the turnover would be AEG with 20% and Auer and Siemens & Halske each with 40%. Whilst the contribution of Auer was much greater than the allocation of shares, including bringing in the name of Osram, Koppel opted for the smaller shareholding for compensations outside the lamp industry. This was the first time in the history of Siemens & Halske in which they did not have the majority shareholding in what was a key part of their business. The third great world lamp company was born. In the following years, under the direction of William Meinhardt (1872-1955), a lawyer and engineer from the Auer Company, Osram acquired many foreign lamp manufacturers in Austria, Poland, Spain and the Scandinavian countries.

GEC, however, continued to have exclusive rights to the Osram name in Britain and the Colonies, through the purchase of the Auer shareholding from the Public Trustee who had acquired it under the Trading with the Enemy Act in 1916. From the formation of Osram GmbH in 1920 this anomaly was to cause unease in the German operation for almost seventy years until Osram GmbH acquired Osram GEC in 1990.

The merging of Electric Lamp Service and Atlas Lamp Works in

1937 was the start of the amalgamations that were later to form the Thorn Empire; after the start of the war (1940), they acquired Vale Lamp Works. Thorn under Atlas started the manufacture of fluorescent lamps in the early 1940s; in 1949 they made a 'know-how' agreement with Sylvania of America. E.K.Cole (brand Ekco) acquired Ensign Lamps in 1946 and began a close working relationship with Thorn, who later acquired E.K.Cole itself.

1. The Electric Lamp Industry – Technological Change & Economic Development from 1800-1847 by A.A.Bright
2. The United States versus General Electric Company and others, Circuit Court of the United States, Northern District, Ohio. In equity No. 8120;1911
3. The Electric Lamp Industry – Technological Change & Economic Development from 1800-1847 by A.A.Bright
4. The Electric Lamp Industry – Technological Change & Economic Development from 1800-1847 by A.A.Bright
5. The Electric Lamp Industry – Technological Change & Economic Development from 1800-1847 by A.A.Bright
   via: Franklin Institute, Incandescent Electric Lamps, 1885; Howell and Schroeder, the History of the Incandescent Lamp 1927; Schroeder, History of Electric Light 1923.
6. The History of Philips Electronics N.V., Vol. 3, by I. J. Blanken (Translated by C. Pettiward)
7. Monopolies and Restrictive Practices Commission – Report on the supply of Electric Lamps, 1951
8. Siemens Brothers by J.D. Scott
9. Command Paper 9072, paragraph 4 .
10. The Times 21st January 1926
11. The Telegraph and Telephone Journal, February 1928, quoted in Committee on Industry, vol. 4
12. Anatomy of a Merger by Robert Jones & Oliver Marriott
13. Anatomy of a Merger by Robert Jones & Oliver Marriott
14. Anatomy of a Merger by Robert Jones & Oliver Marriott
15. Anatomy of a Merger by Robert Jones & Oliver Marriott
16. Anatomy of a Merger by Robert Jones & Oliver Marriott
17. The History of Philips Electronics N.V., Vol. 3, by I. J. Blanken (Translated by C. Pettiward)
18. The History of Philips Electronics N.V., Vol. 3, by I. J. Blanken (Translated by C. Pettiward)
19. A letter from A. F. Philips to G. L. F. Philips, dated 4th July 1922, Philips Corporate Archives
20. The History of Philips Electronics N.V., Vol. 3, by I. J. Blanken (Translated by C. Pettiward)
21. Report of Patent, Design and Trade Mark Cases, Vol. XL11, 5, p2000, 1925

4
Development of Other Lamp Technologies

The early period leading to the development of the incandescent lamp has been covered in the first chapter. The incandescent lamp has been the early industry's 'holy grail' and it has certainly been the focus of the industry for over a hundred years. This was probably due to it becoming the main consumer lamp, being very competitively produced and sold. The other significant influencing factor was the focus on its sensitive consumer pricing leading to government investigations (Monopoly Commissions), which looked into all aspects of the industry, not only in Britain; similar investigations happened in the USA and Europe.

Whilst the incandescent lamp is, electrically, a fairly simple product, tremendous research and development went into the production of the materials and, as previously mentioned, a great deal of chemistry went into the production of the components, i.e. the filament, the lead wires and the glass. In addition to the materials, the glass to metal seals, the gases and construction were all aspects of where competitive advantages were established. So while the incandescent lamp had initially universal appeal, it was always clear that other more efficient solutions would have to be found for many commercial and industrial applications. Indeed, a large variety of lamp developments were created throughout the whole of the 20th century. Indeed, many of these successes were again dependent on other developments, such as an efficient vacuum or the materials needed.

The aim of this chapter is to give a good overview of these other light sources. Inevitably it is necessary to give an appreciation of the technical aspects leading to the development but this is kept to a general level and is not intended to give full technical details of all the different lamps.

The development of arc lamps

While the arc lamp can be run on d.c. (direct current) or a.c. (alternating current), it was originally developed on d.c. Indeed, this was generally preferred, because of the higher arc stability. Fundamentally, an arc lamp consists of two rods of carbon connected to the terminals of a current source. When the carbon rods are brought together and then separated to form a gap of a few millimetres, a brilliant light is created. It is not so much the arc itself that emits the light, but rather the ends of the carbon rods, which are brought to incandescence. The first developers used charcoal rods for their experiments. These burned away very quickly, and the harder retort carbon – a by-product of the gas-works – was soon to be a better alternative.

As with the gas mantle, it took a considerable time before the underlying processes were properly understood. With the d.c. arc, it is the positive carbon tip (anode) that has the highest temperature, and therefore emits the most light. It is also the one which burns away most rapidly.

The chief advantage of the electric arc was its brilliant, highly intense light, which never ceased to amaze 19th-century spectators. Its drawbacks, however, were many. The carbon rods burn away with time and require constant attention to the right spacing, otherwise the arc will extinguish. For the same reasons, the carbon rods (or electrodes) have to be renewed at regular intervals. With d.c. supply, consumption of the positive carbon is about three times as fast as the negative one, assuming they have the same thickness and composition, because of the higher temperature. When burning, a carbon arc hisses and fumes. Even the lowest practicable light intensity is still far too high for domestic use. Irregularities in the arc current, combined with the negative voltage-current characteristic of the arc, for a long time ruled out the possibility of connecting more lamps in parallel to one and the same power supply. In the beginning, each arc lamp even needed its own battery or generator. Later, to a limited extent, series supply became possible.

The 19th-century inventors managed to reduce or eliminate most of these drawbacks, but the arc lamp remained an expensive, temperamental and cumbersome light source. In 1846, William Edwards Staite commenced further experiments and later patented many improvements, most notably a device to maintain the carbon rods at the proper distance during burning by means of an electromagnet through which the lamp current flowed. The magnet kept the carbon rods apart against gravity, but its force would weaken as the carbon rods faded away and the arc lengthened causing the current to eventually fail. The idea was further developed by Foucault and Dubosq and put into practice in France by Serrin in 1859; he found a method of maintaining the position of the arc, despite unequal burning of the positive and negative carbons.[1]

Current regulated lamps of this type could not be connected in series to the current supply, as every regulator in the circuit would respond to the current variation, whichever lamp required regulating. To overcome this problem, the English engineer R.E.B.

Crompton (1845-1940), Brush, and Wallace-Farmer in the USA, devised a system whereby the voltage across the arc was monitored, instead of the arc current, and regulation was performed by a shunt electromagnet of comparatively high resistance. A series electromagnet was still necessary to start the lamp by pulling the carbon rods apart, thus creating the arc.

In 1860 J. T. Way developed a moderately practical arc lamp for the period (Figure 12), which was used to light Hungerford suspension Bridge in London. Way's lamp consisted of a carbon arc enclosed in an atmosphere of air and mercury vapour. This was classified as the first mercury arc lamp.

Finally, around 1880, Crompton and Pochin in England and Friedrich von Hefner-Alteneck in Germany, developed the differential carbon-arc lamp. Here the lamp power was kept constant by monitoring both the arc voltage and the arc current and the regulating mechanism was considerably refined by means of a clockwork escapement.[2]

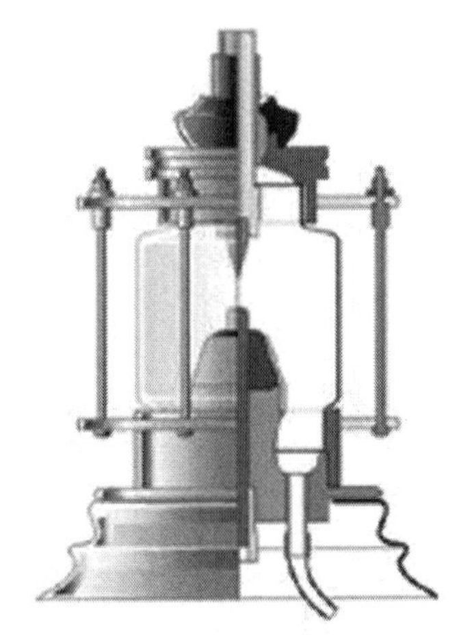

Figure 12: J.T. Way's lamp mercury arc lamp

If the air can enter freely, as is the case with open-arc lamps, carbon rods burn away at an average rate of 20 millimetres per hour. In 1893, William Jandus and Louis B. Marks introduced the enclosed arc, whereby the arc was contained in a glass balloon. This reduced carbon consumption to about one fifth and allowed burning times of up to 150 hours without carbon replacement.

Another important improvement was the flame-arc lamp, invented by Hugo Bremer in 1889. By adding fluorides of certain metals to the carbon rods, the luminous output of the arc could be considerably increased without increasing the electricity consumption. Also, the type of salt added could influence the colour of the light. Typical flame compounds are rare earths for white light, calcium for yellow, strontium for yellow and iron for ultraviolet.

Probably the earliest application of the arc lamp was on the theatre stage in Paris at the première of Giacomo Meyerbeer's grand opera 'Le Prophète' in 1849, in which the arc lamp simulated the sun. This was followed by a period in which no opera or ballet performance was complete without arc-light effects. Although a carbon arc lamp had been demonstrated in Paris as potential for street lighting by Joseph Deleuil in 1844, it was not until 1878 that it was used to practical effect in street lighting, again in Paris. It was not long before many other major cities in Europe followed. Whilst it had no practical application in domestic lighting, it was used in factories, large stores and railway stations.

The heyday of the carbon arc lamp came in the closing years of the19th century, after which it was quickly superseded by the incandescent electric lamp. The main applications remained as those where a concentrated source of extremely high intensity was required. Typical examples included stage lighting, light in photo and film studios, cinema projectors, and searchlights (used by the Army and Navy before aircraft existed. . With the advent of the short arc xenon lamp in 1951, its role in these fields of application has all but ended.

The development of discharge lamps

It is probably accepted that the oldest observations of electric discharges in rarefied gases date back to the 17th century. Beginning in 1676, shortly after Torricelli's invention of the mercury barometer, scientists time and again reported the appearance of light producing phenomena in the vacuum above the mercury. In the following century, experiments with light effects in evacuated glass globes became popular.

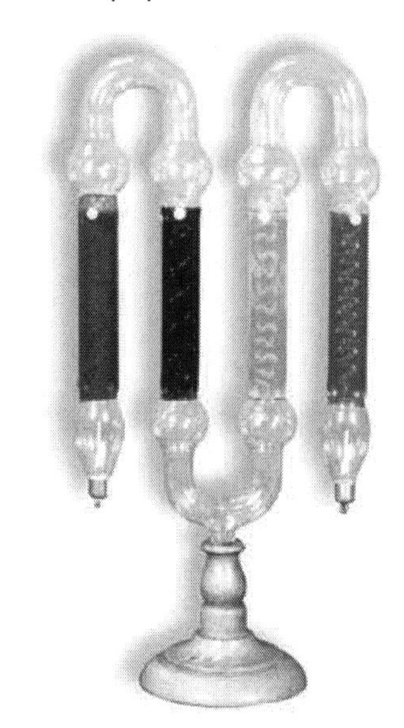

Figure 13: Geisler tubes with discharge through fluorescent fluids

Many scientists developing the carbon arc lamps also later concentrated on gas discharge lamps. Apart from his work on the arc, Davy was one of the pioneers investigating the conduction of electricity in gases. His work (and Faraday's) led to the well- known Geissler tubes (1856), in which he experimented with electric discharges in evacuated glass tubes. He found that at certain pressure, these would emit a bright violet light. Subsequently, investigators like Hittorff, Crookes and Goldstein found that the light phenomena changed by further reducing the pressure in the tubes or by adding other gases to the rarefied air.[3] So while light-producing current flow in low-pressure mercury vapour was known as early as the time of Geissler, again, a long period elapsed before practical light sources were made. Indeed, these Geissler tubes were later used for decorative purposes as shown in Figure 13 (c.1890) in which the discharge was through fluorescent fluids. The low-pressure discharge in mercury vapour is itself of low luminous efficiency and unsatisfactory colour. It has little or no application in this primitive form. Its great value in the now common tubular fluorescent lamp was to appear much later.

As early as 1862, the first British patents were granted to Timothy Morris, Robert Weare and Edward Monckton, who proposed using coloured light from Geissler tubes filled with various gases or vapours in signalling and lighting buoys. In 1866, Adolphe Miroude was also granted a British patent for a battery-operated nitrogen-filled Geissler tube for buoy lighting.[4] Perhaps it was the growing dissatisfaction with the carbon filament lamp and its limited use in commercial applications in the 1885-1900 period that turned scientific and development efforts to seek improved alternatives. One of the early attempts to use the phenomenon of gas discharge for lighting purposes took place at the end of the 19th century. During the early years of the 20th century many different and novel lamps appeared. Suffice it to say that many of these never went into commercial production; however, some of the most successful were discharge lamps with vapour tubes. Some of these began to replace incandescent lamps for special applications.

In 1894, D. McFarlan Moore began experimenting with tubes filled with nitrogen or carbon dioxide at low pressure. The division between *low-pressure* and *high-pressure* discharge lamps is generally regarded as one of atmosphere. With nitrogen, a pinkish light was obtained, while with carbon dioxide, the colour came very close to that of daylight. Initially, Moore used external electrodes in the form of metal foil wound round the tube ends. From 1902, he used internal graphite electrodes. The supply voltage of several kilovolts was obtained from a transformer, and hence these lamps became classified as 'high voltage' lamps. The main disadvantage was the very low output per unit of surface, about one tenth of a modern fluorescent tube. In 1898 D. McFarlan Moore created the first installation in Madison Square Gardens using tubes with external electrodes and filled with carbon dioxide, which was absorbed by the glass tube wall and hence had to be topped up with carbon dioxide from time to time. Nevertheless, the early installations lasted for nearly thirty years because of their good colour matching qualities.

In 1907, working on a similar principle to the Moore lamp, the French physicist Georges Claude (1870-1960) made his first tubes filled with low-pressure neon. His early experiments were aimed at producing a lamp for general lighting purposes. He soon discovered that the brilliant red light was better suited to advertising signs, and that he could change the colour by adding other gases or vapours. From this and other rare gases a wide range of coloured tubular lamps has been developed, mostly applicable to signs and decoration. He originally formed a company in 1902, Société l'Air Liquide, for liquefying air and separating its constituents; the lamps became a natural by-product to use these gases. In 1910, he first demonstrated his neon tubes on the façade of the Grand Palais in Paris and in the Church of St. Ouen at Rouen. In 1923, his younger cousin André Claude joined him, and they formed Société Claude Luminère, to make neon tubes. Neon tubes are widely used today but they are generally given an internal fluorescent coating to increase the light output and provide a greater choice of colours.

Meanwhile, another scientist, Peter Cooper Hewitt, in 1901 demonstrated the first low-pressure mercury lamp employing a one-metre-long tube with an electrode of iron or graphite at one side. It was mounted horizontally and contained enough liquid mercury to make an electrical connection between the electrodes at each end. To turn on the lamp, a voltage was applied to the electrodes and the tube tipped slightly, so the mercury ran to one end and an arc was struck as the mercury connection broke (Figure 14). Cooper Hewitt made many demonstrations of his improved mercury lamps and his work was covered by a series of patents in 1900-1901. He also dealt with the problem of starting the lamp, as tipping the tube containing mercury was not very convenient. He also needed to have a series resistor or an external 'ballast' resistance in the circuit to limit the lamp current.

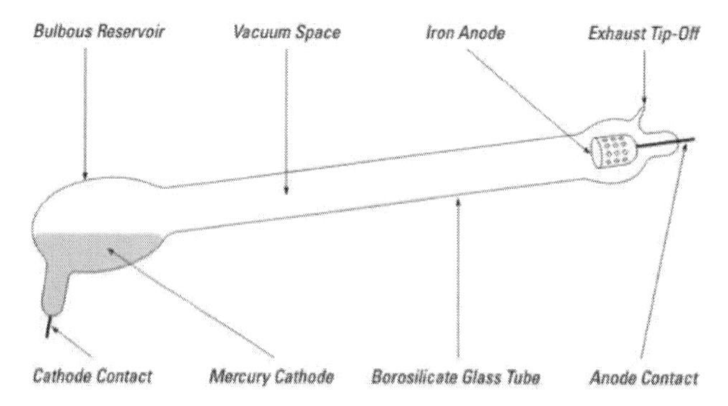

Figure 14: The principle of the Cooper Hewitt lamp

In his many experiments he established that an alternative method of starting was to pre-heat the end of the tube with a Bunsen burner or electric coil. He also introduced a secondary gas through which he passed a current, which would then generate heat and warm the mercury so that it became conducting.

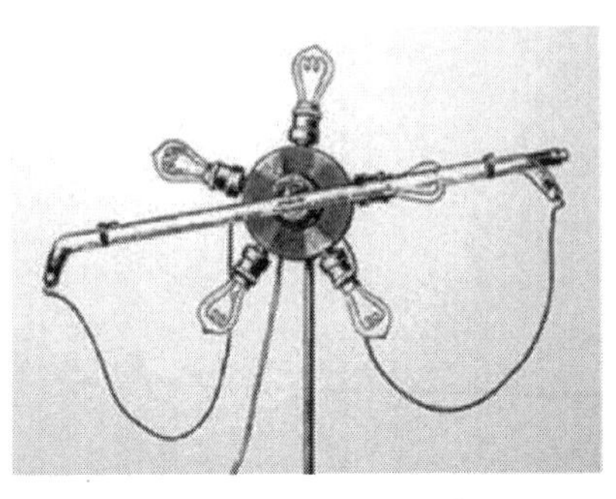

Figure 15: Free-standing Cooper Hewitt lamp with carbon filament lamps in series to act as ballast

The Cooper Hewitt Electric company, financed by George Westinghouse, was established in 1903, and published data showing, surprisingly, 40 lumens/ Watt,[5] which was probably the most efficient lamp of its type at that time. There were, of course, still two key problems to solve; one was the flicker, and the other was the colour, which was a bluish green and contained virtually no red elements. Whilst his company tried to make a feature out of the colour appearance, it proved to have very limited appeal and solutions came only slowly over the next 50 years. However there

Figure 16: Typical advertisement for the Cooper Hewitt lamp in 1908

is little doubt Cooper Hewitt made a significant contribution to the early developments of mercury discharge lamps and his lamp soon became obsolete, although it did have some success in industrial applications and in the photo-printing business. It was to re-appear in the late 1920s, much improved as a fluorescent lamp.

The improvement in the colour rendering of the low-pressure mercury lamp came from two directions, first from the use of fluorescent coating and secondly from increasing the mercury-vapour pressure. R. Küch and T. Retschinsky developed the first high-pressure mercury lamp (Figure 17) in 1906. Because of the high temperature and the pressure inside the discharge tube, it was made of quartz. This basic concept was built into a lamp and marketed under the name of 'Quartzlite' by Brush Electrical in 1908. The success of this lamp was only modest; the main problem was that quartz readily transmits ultraviolet radiation, which was still present in considerable quantities.

Also, mass production of the lamp was problematic due to the unsatisfactory solution found for making a hermetic seal between the lead-in wires and the quartz discharge tube. Interest in the high-pressure mercury lamp also therefore soon faded, to be re-awakened in the early 1930s, when several companies came out with improved versions. One of the early lamps to be marketed was by GEC under the name of 'Osira'. The lamp used a discharge

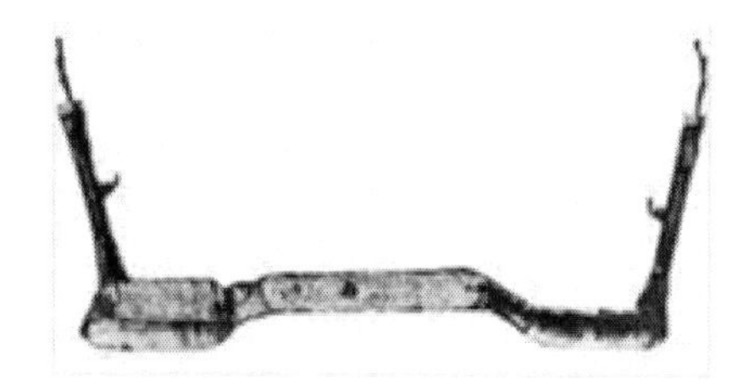

Figure 17: Küch and Retschinsky's quartz mercury arc lamp 1860

tube made of aluminosilicate hard glass, but again the sealing techniques between the quartz and tungsten were unsatisfactory. Initially this lamp was restricted to burning in the vertical position, but this was later solved by the use of an electromagnet that kept the arc straight when the lamp was horizontally burned. Nor had the early lamps an integrated starting aid like an auxiliary probe. To solve this, GEC fitted each luminaire with a small Tesla coil in order to ignite the lamp. This was the first time an ignitor was used

Figure 18: The first high-pressure mercury quartz lamp in 1936 (Philips HP300)

These had a discharge tube of hard glass, instead of quartz, so avoiding the sealing problems. However, in 1935 Philips found a way to seal tungsten wires in quartz. The pressure in the discharge tube was increased from just over one atmosphere to ten, thus achieving a far better spectral distribution of the emitted light. This enabled Philips to market the first low power high-pressure lamp (Figure 18). A fluorescent powder that could withstand the intense ultraviolet radiation was discovered in 1934 in the form of cadmium sulphide.

The corrective effect on colour was only moderate and it was only in the post-1945 period that several other fluorescent powders were tried for improved colour. The introduction of the colour-corrected mercury lamp was made possible with the development of the manganese-activated magnesium germinate and fluorogerminate in 1950. This greatly improved the colour-rendering index and had a beneficial effect on lamp efficacy. In

1967, yttrium phosphate vanadate, a derivative from TV development technology, was introduced to eventually become the most popular; it is still in use today.

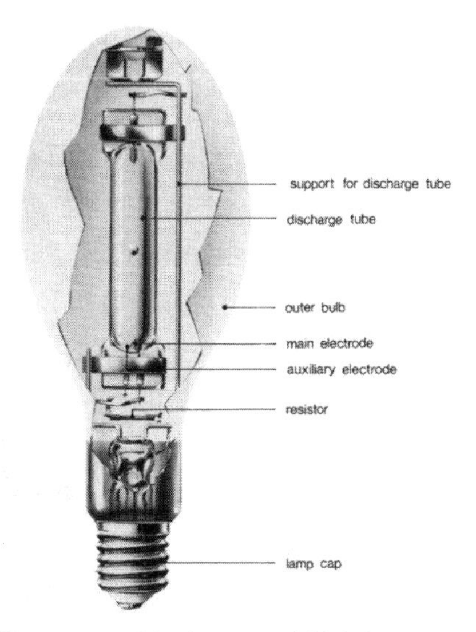

Figure 19: Modern ovoid high-pressure mercury lamp

There were two further offspring of the high-pressure mercury lamp to be noted, the first being the *mercury blended lamp*. This was a combination of an incandescent filament and a high-pressure mercury discharge tube, connected in series and mounted in a standard bulb. The filament acts both as a current stabiliser and provides some colour correction towards the red end of the spectrum. This can be classified as a direct derivative of the Cooper Hewitt lamp and it was first manufactured in 1935.

The second offspring came from a patent taken out by G.H. Reiling, in 1961, on a high-pressure mercury lamp that had

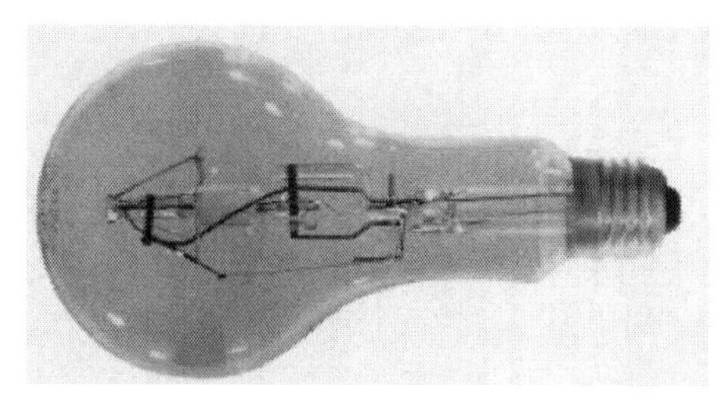

Figure 20: Early Philips Blended lamp 1941

halogen compounds of certain metals added to the mercury-vapour filling. These compounds were typically: indium, thallium and sodium or scandium and sodium or dysprosium and thallium. This became known as a metal halide lamp and it was marketed in 1964. It was a significant improvement over the traditional high-pressure mercury lamp, both in terms of increased efficacy and better colour rendering.[6]

While mercury was chosen for metal-vapour discharge lamps, sodium was also promising, as it emitted a large proportion of its radiation in the visible part of the spectrum, particularly the yellow part of the spectrum which is very close to the region of maximum eye sensitivity. Experiments in 1922 by M. Pirani and E. Lax in Germany and in 1923 by A.H. Compton and C.C. van Voorhis in America showed that that very high efficacy could be achieved. Philips and Osram constructed the first practical low-pressure sodium lamps in 1931, and then the efficacy was 50 lumens/watt. Today that has been increased to 200 lm/W. In 1964 Bill London and Kurt Schmidt from the USA successfully produced high-pressure sodium lamps resulting in dramatically improved colour characteristics but at the expense of efficacy. High-pressure sodium lamps continue to be developed and through the introduction of White SON in the 1980s gave further improvements in both efficacy and colour.

As already mentioned, all discharge lamps need a starter (or ignitor) to create the arc and a ballast to stabilise the arc, commonly referred to as control gear. While some of the larger lamp manufacturers developed their own gear, there were also a number of specialist companies producing this control gear, widening the range of manufacturers within the industry.

The Development of the Fluorescent Lamp (Low pressure mercury discharge)

It was not until 1926, when German scientists Friedrich Meyer, Hans Spanner and Edmund Germer described how electrodes could be preheated to facilitate ignition at low voltages, that the fluorescent lamp could be produced. They published a report describing how the electrodes could be pre-heated to facilitate ignition at low voltages and how, with the tube wall coated with a fluorescent material, the strong ultraviolet radiation of a mercury discharge could be converted into visible light, creating the early fluorescent lamp.

The colour of the early fluorescent tubes was very unsatisfactory; indeed it was several years before the first hot cathode fluorescent lamps were demonstrated publicly at the Annual Convention of the Illuminating Engineering Society of North America in 1935. This resulted from the work of André Claude, the cousin of the inventor of the neon lamp Georges Claude, who took out a patent for the hot cathode fluorescent lamp in 1932, and his patent rights had been taken over by GE. The first installations were in 1936 in America; Osram quickly followed these with demonstrations at the World Exhibition in Paris in the same year. In the next two years several major European lamp manufacturers started to make fluorescent lamps, but this came to an abrupt end with the start of the 1939-45 war.

The early phosphors were calcium tungstate and zinc silicate giving an efficacy of approximately 30lm/W. Some development continued in Britain, and in 1942 A. H. McKeag of GEC discovered that activated halophosphates of calcium and strontium had very good fluorescent properties, and these were introduced in 1946.

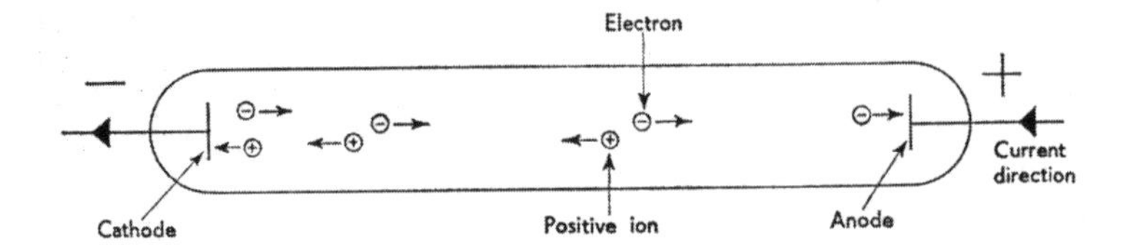

Figure 21: Electric discharge through a tube of ionised gas

These almost doubled the luminous efficacy; however, although huge improvements were made in the efficacy, the colour appearance remained poor resulting in limited applications where colour appearance was less important the efficiency of the lamp.

At this stage it is worth mentioning that all discharge lamps require an electrical control circuit. In its simplest form, when a current is applied there is a rapid increase in current after initiation of the discharge and this would continue until the lamp failed or the fuse blew. Hence the gas discharge/current is stabilised by the introduction of a ballast in series with the lamp. If the same potential difference, which is required across the electrodes to maintain the discharge under stable conditions, is applied across the discharge path when the discharge is not operating, then usually nothing will happen. In order to start the discharge it is necessary to apply a higher voltage (the difference between this and the operating arc voltage remaining across the ballast), in addition to heating the cathode or temporarily applying a higher potential to or in the vicinity of one of the electrodes, or both. In selecting the starting aid to be employed, the composition of the gas, the gas pressure, distance between the electrodes and the tube diameter must all be taken into account.[7] Hence all discharge, including fluorescent lamps, which is a form of discharge lamp, need a ballast to stabilise the current through the lamp and a starter switch to ignite the discharge.

Over the years tremendous improvements have been made in phosphor technology, of which one of the most notable has been the introduction of the tri-phosphor coating by Philips in 1973; this was a spin-off from the TV cathode ray tube developments. This development eventually boosted the efficacy up to 90lm/W, with excellent colour rendering. Previously, good colour rendering had always been at the expense of efficacy. These phosphors allowed the developers to increase the wall power loading within the tube with the resulting reduction of the diameter from 38mm (T12) to 26mm (T8).

In addition. the developments in control gear, starter switches and ballasts, required for all discharge lamps, have also made huge improvements and today this has led to the use of electronic control gear, which has also had significant benefits in fluorescent lamp efficacy. This electronic control gear combines the current stabilising and the starter requirements in one physical unit.

Halogen lamps

It is perhaps surprising that in spite of the tremendous development effort that went into improving the incandescent lamp, it was not until 1959 that two development engineers, Edward Zubler and Frederick Mosby, at GE of America discovered and published a paper showing a tungsten halogen lamp with nearly 100% lumen maintenance, an approximately 20% increase in efficiency and considerable increase in life. This was in a tubular lamp much smaller than ordinary incandescent lamps.

Surprisingly, GE regarded this technology as only applicable for niche applications. While it was much smaller, it was expensive to produce, and seen as applicable for projection applications where conventional incandescent lamps were normally too large; and even then the filament was too large to create a point source from which an image could be projected.

It is worth examining the technology behind the halogen cycle. In the conventional incandescent lamps it is not possible to utilise the effect of pressure on filament life because the bulb is too frail to increase the fill gas pressure much above atmospheric pressure. To make a strong bulb that can withstand operating pressures in excess of ten atmospheres, it has to be made of much thicker glass or quartz and made much smaller. In the conventional incandescent lamp, the process of evaporated tungsten would cause the inside of a much smaller bulb to blacken more densely and too quickly, causing an unacceptable reduction in efficiency.

Zubler and Mosby discovered that by adding iodine to the gas, the chemical reaction would clean tungsten from the sides of the bulb; this became known as the halogen cycle. Subsequently many engineers have experimented with different combinations of halogens which consists of fluorine, chlorine, bromine, iodine and

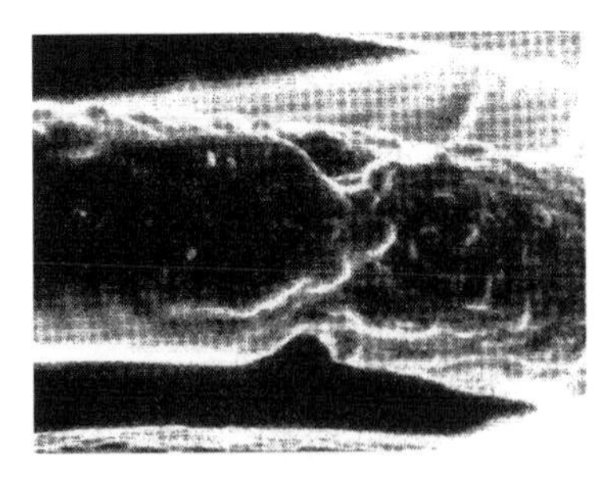

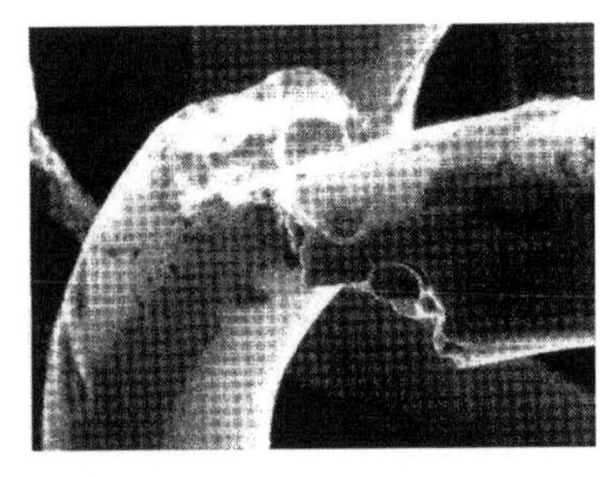

Figure 22: The weak/hot spot in a filament which eventually results in fracture

astatine. In addition to these chemicals, various amounts of hydrogen and even oxygen were introduced to optimise the halogen cycle. The process of tungsten evaporation was well understood.

Tungsten always evaporates from the hot spots on the filament; in the conventional incandescent lamp, this usually deposits itself on the inside wall of the glass envelope causing blackening, and these hot spots get hotter the thinner the tungsten wire gets and the process accelerates. With the halogen cycle the evaporated tungsten deposits itself back on the wire but the real secret was to get the tungsten to deposit itself back on the hot spots from which it came in the first place. This has never been fully achieved but great strides have been made in the refining of halogen lamps including the internal infra-red coating on the inside wall of the lamp to reflect the heat back on the coil.

Initially, halogen lamps were used for specialist projection equipment and in particular for low voltage applications, where the low voltage filament coil was small, enabling a reflector to create a good projection. Indeed one of the main initial applications was for motor-car headlights and spotlights.

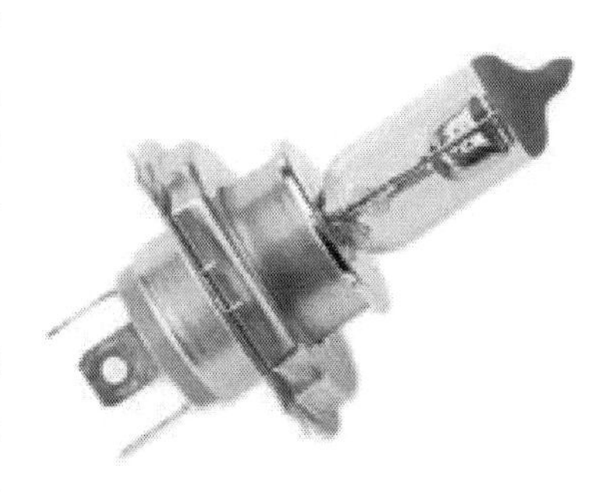

Figure 23: H4 Halogen car headlight

Alongside these developments were halogen lamps for the specialist applications in stage, theatre and film. Until the development of halogen lamps, these had been very large incandescent lamps. Later, linear halogen became important for floodlighting in both professional and domestic markets.

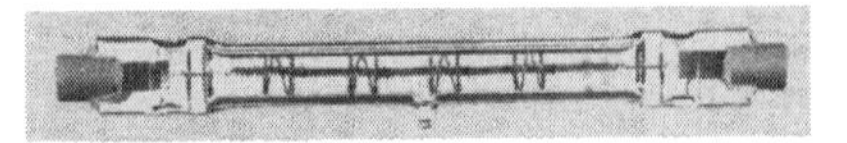

Figure 24: Linear halogen lamp

Figure 25: Halogen burner mounted in an incandescent lamp

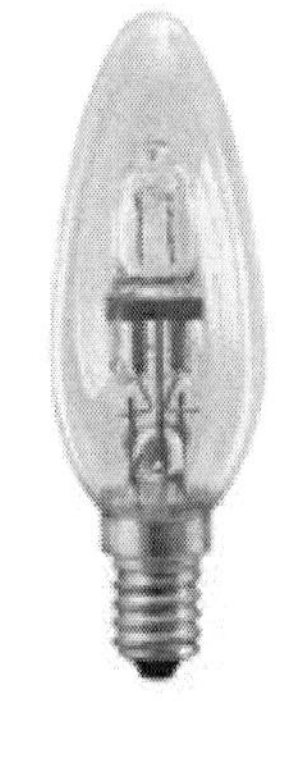

Figure 26: Halogen burner in a candle lamp

High-efficiency halogen lamps are being enclosed in conventional outer envelopes to replace incandescent lamps as the legislation bans these conventional lamps from the market. These halogen lamps are between 20-30% more efficient.

These lamps can be dimmed and can replace conventional incandescent lamps, but, as the burner is usually filled with a gas at a pressure of several atmospheres and reaches very high temperatures, it has to be enclosed in another build to ensure it is used safely.

Compact Fluorescent Lamps (CFL)

Figure 27: An early Philips compact flourescent lamp-SL 1976

To the technicians the development of compact fluorescent lamps was not a completely new concept. But with the advent of the 'oil crisis' in 1971 and the dramatic rise in the cost of energy there was a new drive to seek alternative more efficient light sources, in particular for the incandescent lamp.

Although not fully appreciated at the time, this development race was about to create a new era in lamp and lighting. As so often happens, there were a number of separate development initiatives, which came together in the 1970s probably coinciding with the post-'oil crisis' market conditions. The main factors influencing the lighting industry were a much greater awareness of the cost of energy, the fact that energy sources were not unlimited, and the concern for the environment.

With the advent of the new generation of tri-phosphor technology, the result was a fluorescent lamp with a much more acceptable output in the home or domestic environments. The colour rendering was coming nearer to that of the incandescent lamp; the trick was to try and create a replacement light source for the incandescent lamp. Philips was the first to take the tri-phosphor technology, reduced diameter tube and bend it into a 'double U' to reduce it sufficiently in size to enable it to be encapsulated in a glass envelope. It was then combined with a starter switch and a ballast in a base with a conventional cap (bayonet-BC or screw-ES). This became known as the SL lamp.

Philips protected this development with patents and tried to license other manufacturers. This was not the success originally envisaged; the lamp combined with the conventional gear was seen as rather clumsy and instead of the main market being the home, it was seen as a huge energy- saver in the industrial and commercial markets where incandescent lamps were still in use. In this particular development America was behind Europe, probably because the drive to energy-saving was not as great in the initial period.

The immediate successors of this SL lamp were more simple bent fluorescent tube (16mm) in a 'U' shape with a newly introduced base (cap), leaving the gear to be incorporated within the luminaire (lighting fitting). Philips and Osram combined, although with different construction, using the same base and wattages, which were hence interchangeable. Thorn produced a format called the 2-D lamp as the bent fluorescent tube was shaped in the format of two D's facing each other. All these required the luminaire manufacturers to produce new luminaires; this meant that penetration into the market was slow.

A solution was about to emerge. As previously outlined, all discharge light sources require some form of control gear; the development of electronics dedicated to a particular lamp technology proved to do the job much more efficiently than the old conventional control gear. With the ability to miniaturise the electronic circuit using an integrated chip, the original concept of an integrated compact fluorescent lamp (CFL) with the compacted fluorescent tube and the control gear was now more of a reality. Various concepts were developed, these being much lighter, brighter (higher efficacies) and at more acceptable consumer prices; hence, the volumes increased dramatically.

Today these are accepted as part of the package of lamps for industry and commerce and, reluctantly, for the home.

The Development of Low Pressure Sodium Lamps (SOX)

Mercury had been the obvious choice for a filling gas for metal vapour discharge lamps, because, of all the metals, it is the only one with appreciable vapour pressure at normal temperatures. It also emits a reasonable proportion of its radiation in the visible part of the spectrum. However, sodium, which has a melting point of only 96°C, emits almost all of its radiation in two closely separated lines in the yellow part of the spectrum. This is very close to the region of maximum eye sensitivity. In theory, this principle of a sodium discharge should lead to a very efficient lamp.

In 1922, M. Pirani and E. Lax of Osram in Germany carried out the first experiments with low-pressure sodium discharges. A. H. Compton and C. C. Vourhis followed this in 1923 with further experiments in the United States, showing very clearly the very high efficacy that could be achieved. Fortunately, in 1920, A. H. Compton had discovered a sodium resistant borate glass, which was essential for the discharge tube. Alkalis, being string reducers, require special glasses as normal materials like soda-lime silicates are readily attacked and lead to the formation of a brown light-absorbing film.

Both Philips and Osram pursued this technology, and produced the first practicable working lamps in 1931. This was quickly followed by the first road lighting installation in 1932, using a d.c. supply, in the south of Holland. The a.c. supply version was designed in 1933 and Philips installed the first UK installation in Croydon later that year. The heat generated by the low-pressure sodium discharge was barely enough to keep the lamp at an optimum operating temperature, so the early lamps had the discharge tube encapsulated in a double walled evacuated glass mantle. In 1955 a single evacuated tube coated with a heat reflecting material, initially tin oxide, was used, and nowadays this is indium oxide. This light source became the most efficient in common use with efficacies up to 200 lm/w, which is about fifteen times that of an incandescent lamp. In spite of its very high efficiency, this lamp is virtually monochromatic, operating only in the yellow part of the visible spectrum, making it almost impossible to identify any other colours. This confined its applications mainly to street lighting and security lighting where colour definition is less important.

Figure 28: The first Philips d.c. low-pressure sodium lamp with a vacuum mantle

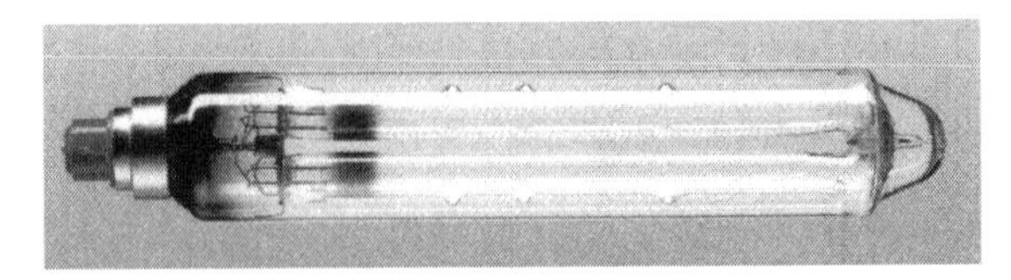

Figure 29: A modern low-pressure lamp with a U-shaped discharge tube in an evacuated clear outer tube

The Development of High Pressure Sodium Lamps (SON)

Over the years, many attempts were made to improve the colour characteristics by increasing the vapour pressure, which also reduces the efficacy; unfortunately it was then found that the highly aggressive sodium vapour would attack all types of glass including quartz. In 1955, R. L. Coble discovered that a gas-tight, sodium-resistant, translucent discharge tube could be made from highly purified, sintered aluminium oxide. Because of the difficulties in managing the production process, it was nearly ten years before Bill London and Kurt Schmidt of the USA produced practicable high-pressure sodium lamps. In Britain, Thorn Lighting was probably the first to produce this lamp technology in 1967.

In the modern lamp the discharge tube contains an excess of sodium to give saturated vapour conditions when the lamp is running. An excess of mercury is also present to provide a buffer gas and xenon is included, to facilitate ignition and limit heat conduction from the discharge arc to the tube wall. The discharge tube is housed in an evacuated protective glass envelope.

The next significant improvement came during the 1980s with the ceramic metal halide lamps referred to as 'White SON'. These gave a colour rendering much nearer to incandescent than any previous development.

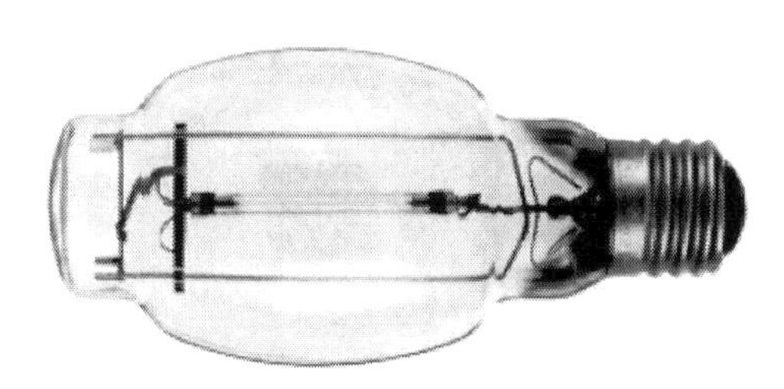

Figure 30: An early Westinghouse high-pressure sodium lamp with a clear sintered aluminium oxide

During the 1990s considerable efforts were made by Philips to make a lamp more environmentally friendly by the removal of mercury from these lamps. Significant progress was made by increasing the xenon pressure and introducing some additional starting aid such as sintered metal strips on the discharge tube surface.

Because these high-pressure sodium lamps had such improved colour rendering properties compared to low-pressure, they radiate energy across a good part of the visible spectrum. They soon gained popularity for a variety of applications where colour recognition was important. These included sports halls and arenas, outdoor residential areas; but low pressure sodium, because of its very high efficiency, continued for main road lighting and many security lighting applications.

Metal Halide Lamps

Perhaps of all the lamp categories, the metal halide development followed a typical development pattern. When the mercury lamp was invented in the early part of the 20th century, it was recognised that it lacked red light in its spectral emission even though it was the foundation for the metal halide lamp. However, it had to wait almost sixty years before technology caught up to solve the problems in achieving a much superior type to the basic mercury lamp.

A patent was granted to Charles Proteus Steinmetz on 7th May 1912, to improve the colour of the early mercury lamps by adding halide salts. His principle was to use external 'pools' of mercury contained in an inverted U-shaped tube as electrodes (labelled 'D' in patent diagram) and these were coated with a layer of metallic halides on the surface of the pools (labelled 'F'). Unfortunately, his design left the electrical arc dancing around the surface of the pool, preventing a consistent colour from being generated.[8]

The copy of the patent shown is on display in the National Museum of American History and the hand-written note on the top left, "1st Action", is by G. Reiling, the GE physicist who filed his own patent in 1961 for the modern metal halide lamp. There were many studies in the years prior to the Reiling patent investigating metal-halogen compounds that have higher vapour pressures than metals at a given temperature. It is Reiling, who, today, is generally given the credit for the invention of metal halide lamp, but there was considerable discussion with the American patent office before the Reiling patent was finally granted in 1966.

Originally it was thought that this development would replace most high-pressure mercury lamps. For the technically minded this lamp had a filling of primarily mercury, thallium and sodium iodide that contributed to the sizeable increase in lamp efficacy.

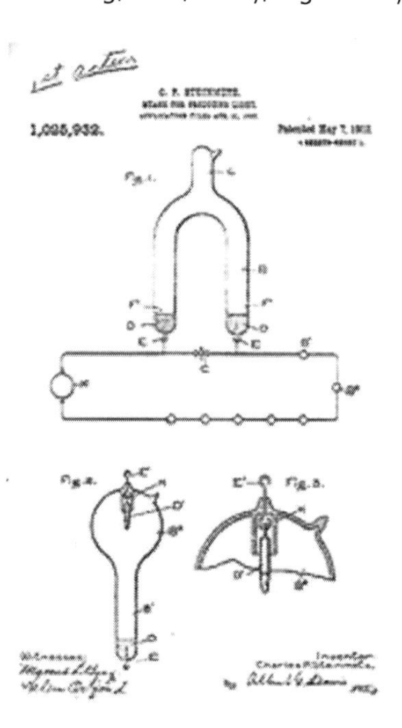

Figure 31: Charles Steinmetz's US patent

During the next 25 years,

considerable competitive development took place between all the major manufacturers both in America and Europe. From being mainly high wattage lamps with high lumen output, these came to be relatively low wattage lamps by the 1980s. Indeed, Philips produced a 35 watt lamp for car headlights but this and other low wattage types had major circuit problems, particularly with a hot re-strike and the cost made them uncompetitive. Eventually breakthroughs began; one of the first was by Thorn Lighting in 1981 as a result of some ten years research; development presented the first metal halide lamp with a sintered alumina discharge tube. Unfortunately, this never reached the market because there was no available ballast in production. In the same year, Osram introduced their compact double-ended HQI-TS for commercial applications, in particular for display lighting in shops.

By 1991 Philips joined by Osram and several car lighting set-makers, in a project they called 'Vedilis' to produce the xenon-metal halide lamps for automotive headlights.

Philips continued with the development of metal halide lamps with a ceramic discharge tube (taken from the development of high-pressure sodium lamps) and in 1995 they launched its range of CDM lamps, soon to be followed by Osram and GE. These

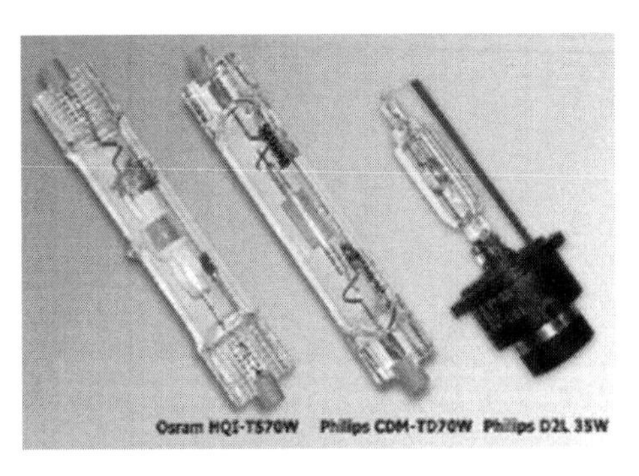

Figure 32: The new lower-wattage metal halide lamps

became known as compact metal halide lamps which offered an alternative to high-pressure sodium lamps with better colour matching, higher efficacies and better colour rendering.

Electrodeless Discharge Lamps

One of the oldest observations of luminous discharges were those taking place in the evacuated space above the mercury in a Torricelli barometer. A fluorescent or neon lamp will light up vividly when placed in a magnetron oven, without being connected to a power supply; the microwave field then acts as the power source. This demonstrates that an electric discharge can be produced without a current being sent between two electrodes.

The electrodes are the most vulnerable parts of any discharge lamp. During lamp life they wear down or lose their emissive power by the impact of fast ions or chemical reactions with aggressive vapours in the discharge tube. Therefore it has long been recognised that there is much to be gained in designing an electrodeless lamp. The most obvious answer to the question of how to initiate a discharge in such a lamp is by the introduction of a high-frequency electromagnetic field from outside, without galvanic contact. There are two ways of doing this: the discharge vessel can be placed in an adapted magnetron oven; alternatively, an inductive coil (or a capacitor) can be introduced into the

Figure 33: Philips electrodeless induction lamp

discharge vessel and connected to a miniature radio-wave transmitter in the base, to act as an aerial.

In the 1980s Philips was the first to introduce their 'Induction' lamp, in which the source of the discharge originates from an electromagnetic field generated by an induction coil antenna (Figure 33). The suppression of the electrodes increases the lamp life by up to sixty thousand hours.

These lamps have a high initial cost but, due to their very long life, they have proved to be economic in difficult and inaccessible lighting installations, where the cost of changing the lamp is very costly. Perhaps scaffolding and/or the closure of a public building would be involved to enable the lamp change.

Special Lamp Technologies

Over the years many special lamps have been developed, and it is not intended to cover them all but to note that many special lamps have been created for special applications. The very nature of light itself, being made from a part of the electro-magnetic spectrum covering ultraviolet (UV) at one end and infra-red (IR) at the other, means that these type of lamps can be used for a huge variety of applications.

The basic discharge along a fluorescent tube without the phosphor coating produces virtually no visible light, but a huge amount of short-wave ultraviolet radiation. Even greater intensity of UV can be obtained with the higher wattage discharge lamps. In fluorescent lamps, this UV can be controlled by the use of phosphors, varying from short wave to long wave, also gradually giving more light, though these lamps are mostly used for their UV radiation, not their visible radiation.

The effects of UV radiation have been studied and classified into three different wavebands within the electro-magnetic spectrum and measured in nanometres (nm):

UV-A (long wave) 315 - 400 nm
UV-B (medium wave) 280 - 315 nm
UV-C (short wave) 100 - 280 nm

Some typical applications for these UV lamps include photocopying, special light effects, detection, curing and hardening of coatings, etching of printed circuit boards and integrated circuits, disinfection, water purification, insect traps, artificial sun-tanning, and so on.

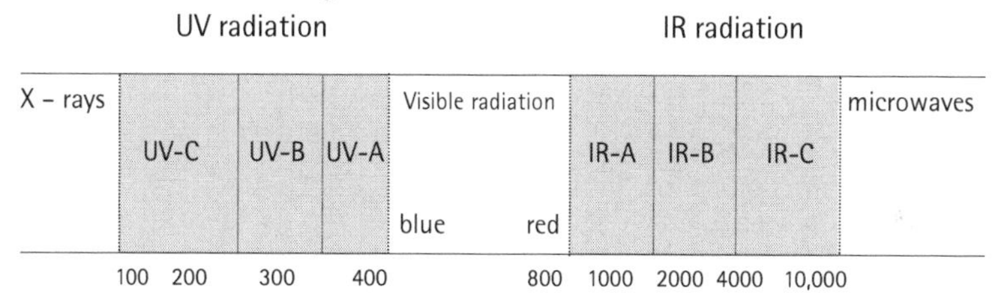

Wavelength in nanometres (nm)

Figure 34: Electromagnetic spectrum showing the sub-divisions of ultraviolet, visible and infra-red wavebands

The effects of ultraviolet radiation fall into two main categories: excitation of fluorescence and initiation of photochemical reactions. Within this seemingly limited scope there is an almost endless variety of laboratory, industrial, commercial, medical, entertainment and domestic applications.

The physical laws governing the generation of ultraviolet

radiation, its propagation and use, and the constructional characteristics of UV radiators and their accessories, are closely related to those governing lighting technology. Just as all the application spheres open to UV are too numerous to mention, so too are the enormous variety of UV sources available, especially with respect to power ratings, which vary from a few watts to several kilowatts.[9]

At the other end of the visible spectrum is the infra-red (IR) and unlike UV, this can be created by an incandescent light source. As with UV, the IR spectrum is divided into three wavebands.

IR-A (short wave)	800 – 1400 nm (0.8 – 1.4 μ)
IR-B (medium wave)	1400 – 3000 nm (1.4 – 3 μ)
IR-C (long wave)	3000 – 10,000 nm (3 - 10μ)

In practice, infra-red wavelengths are often expressed in micrometres or microns (μ), rather than nanometres; one micron equals 1000nm.

Because heating up the filament wire gives out heat, the higher the temperature the greater percentage of the irradiated energy is in the visible or IR-A waveband. The best analogy is to visualise an electric fire or an electric plate on a hob as long wave infra-red (IR-C), which gives out a lot of heat and very little light (approximately 0.2%). At the other end of the scale is the ordinary incandescent lamp, which converts more energy (approximately 6%) into light and less into heat (IR-A). This is the reason the incandescent lamp is an inefficient light source. It is actually a better emitter of heat than of light.

Infra-red lamps have a large number of applications. These include cooking, zone heating, drying of inks, curing of various coatings/adhesives/resins, paint drying, and shrink-wrapping of polyethylene foil. In addition, these lamps have been widely used in personal care for treatment of muscle strain/pain, including rheumatic and back pain. They have also proved to be of great value in the caring of animals, keeping tropical animals warm in cooler climates, the rearing of piglets, poultry chicks, and other young animals. In recent years they have had wide use in hobs for cooking, where they have been able to provide heat much more quickly than the conventional electric iron plates.

The thermal effects caused by infra-red radiation can be simply labelled as radiant heat. Compared with other heat sources, electric infra-red radiators offer high efficiency, fast heat transfer, accurate wavelength control and flexibility, with compactness and cleanliness.

The manufacturers of infra-red heat sources vary widely in construction and working. This is less so with short-wave infra-red radiators, because electricity forms the only practical source of power. Hence this provides a direct link with incandescent lamps where the same operating conditions prevail and similar problems have to be overcome. The transmission and reflection characteristics of short-wave infra-red radiation also strongly resemble those of light, so that the same techniques as in lighting technology can be applied to obtain the exact beam control.[10]

Chronological Development of Lamp Technologies

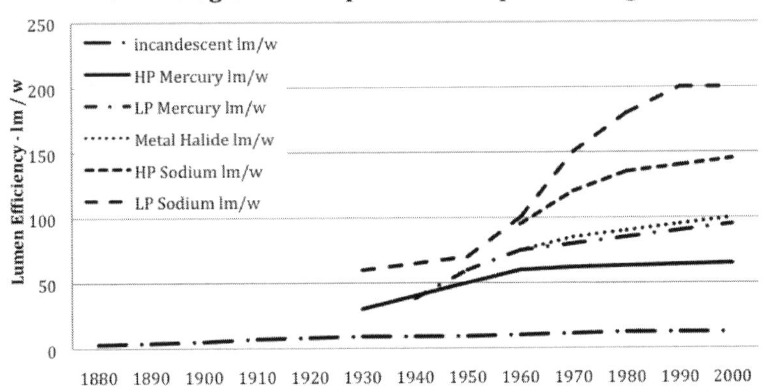

Figure 35: Evolution of developments on electric lamp efficacy over time

Some of the key developments from the above graph influencing commercial production are listed below by lamp type:

Incandescent Lamp

1879
Thomas Edison and Joseph Swan independently succeed in making the first commercially viable incandescent lamps.

1883
Joseph Swan improves the manufacturing procedures of carbon filaments.

1907
Commercialisation of the first squirted tungsten filament lamp was made possible because of the work of Dr. Hans Kuzel.

1911
Vacuum ductile tungsten filament lamps are marketed.

1913
Gas-filled, coiled tungsten filament lamps are introduced.

1933
Coiled-coil tungsten filament lamps are sold for the first time.

1962
Following the research of Mosby and Zuber, GE introduce the tungsten halogen lamp

Low Pressure Mercury

1938
Most manufacturers introduce the low-pressure mercury fluorescent tube.

1946
Fluorescent tubes with halophosphate coatings are marketed.

1961
Osram introduces a fluorescent tube using indium-mercury amalgam to allow the operation of the lamps at higher temperatures.

1973
Philips developed tri-phosphor (and five-phosphor) lamps leading to higher efficacies and better colour rendering.

High Pressure Mercury

1932
GEC introduces its Osira atmospheric pressure mercury lamp, shortly followed by the other major lamp manufacturers.

1936
Philips invents the tungsten-to-quartz graded seal that led to the HP300 and SP500 operating respectively at 20 and 80 atmospheres. These were the first high pressure and super high-pressure lamps.

1951
The first phosphor coated high-pressure mercury lamps appear on the market. The fluorescent powder converts part of the ultraviolet (UV) radiation into red, thus improving the quality of light emitted by the lamp.

1967
Europium activated vandate and phosphor vandate, are introduced as high-pressure mercury lamp phosphors. This material is still used in modern lamps.

Low Pressure Sodium

1932
Philips and Osram introduce the first low-pressure sodium lamps with an efficacy of about 50lm/W.

1933
The first AC positive column types of lamps appear on the market.

1955
Philips introduces low-pressure sodium lamps with a bamboo-formed discharge tube of better sodium resistant glass.

1958
Philips introduces integral low-pressure sodium lamps with the discharge tube enclosed within an evacuated vessel to ensure better lumen maintenance throughout life.

1961
Lamps with discharge tubes enclosed with glass are marketed giving improved thermal insulation to raise lamp efficacy.

1965
Philips introduces a tin oxide infra-red mirror. This semiconductor is spray-coated in the form of a thin film on the inner surface of the outer jacket.

1974
Tin-doped indium oxide film replaces the tin oxide on the inner surface of the outer bulb, which exhibits a stronger infra-red reflectivity, whilst being highly transparent to sodium light.

1983
Philips introduces the SOX-E lamp with improved thermal insulation with the 131W lamp achieving 200lm/W.

High Pressure Sodium

1966
GE introduces the high-pressure sodium lamp under the *Lucalux* brand name. This was made possible by the development of alkali-resistant alumina discharge tube.

ca.1980
High-pressure sodium lamps under the *Lucalox* brand name giving longer lifetime, higher efficacy and better lumen maintenance.

ca. 1995
The first mercury-free high-pressure sodium lamps are introduced. Higher xenon pressure compensates for the absence of mercury in the buffer gas. These usually need additional starting aid in the lamp circuit.

Metal Halide

1964
GE introduces the metal halide lamp. This has a sodium-scandium source designed to replace the high-pressure mercury.

1995
Philips introduces a successful range of CDM metal halide lamps having a ceramic arc tube. This is not a new technology, but new materials and a new technique permit higher operating temperatures, giving higher efficacy and better colour rendering.

[1] History of Light and Lighting by G.W. Stoer, published by Philips Lighting
[2] History of Light and Lighting by G.W. Stoer, published by Philips Lighting
[3] History of Light and Lighting by G.W. Stoer, published by Philips Lighting
[4] The Electric Lamp Industry-Technological change & Economic Development 1800-1847 by A.A.Bright
[5] Lengthening the Day by Brian Bowers
[6] History of Light and Lighting by G.W. Stoer, published by Philips Lighting
[7] High Pressure Mercury Vapour Lamps and their Applications by W. Elenbaas
[8] Paper on 'Lighting A Revolution' American history
[9] UV Radiators and Applications by Philips Lighting.
[10] IR Radiators and Applications by Philips Lighting.

5
Trade Associations

Early Trade Associations

In 1894, the two leading producers in Germany, Siemens & Halske and Allgemeine Elektrizitäts Gesellschaft (A.E.G.), met with the objective of raising standards and stabilising prices; this could be regarded as one of the earliest lamp rings. This was the precursor to a much more comprehensive and international agreement of carbon filament lamp manufacturers *'Verkaufstelle Vereinigter Glühlampenfabrieken'* VVG (Association of United Incandescent Lamp Manufacturers) which was signed in Berlin in 1903 by the main producers in Germany, Austria, Hungary, Holland, Switzerland and Italy. It consisted of twenty leading incandescent lamp factories in continental Europe. The highly successful partnership of Gerard and Anton Philips was achieving remarkable success, with Gerard as the engineer/developer and Anton as a wily sales/businessman, becoming a significant force in the market. Anton regarded it as a triumph to be invited to be a founder member of this international cartel. An example of how Anton operated is shown in this passage from Dr. Bouman's book on Philips:

> Anton met the Berlin directors of the AEG and was advised to cease his efforts to compete with them. 'I have only to press this button' one of them said – it was the great Emil Rathenau himself (founder of AEG) – 'and the price of lamps will go down to twenty pfennig each.' Anton replied 'you may press as many buttons as you like, but you will always find yourself too late.

> All our contracts of last month contain the clause that the price of Philips lamps will be half a pfennig below the AEG price, come what may.'

Philips then flooded Berlin, the head office of AEG, with cut-price lamps. This was typical of actions that will have speeded up negotiations to set up the international Association. Because of this extreme price competition, a mutual price regulation was established and over-production had proved very unprofitable. The Association therefore fixed better prices, but at the same time, in order to prevent over-production, each company was fixed to a previously agreed quantity, i.e. a fixed quota of lamps. Financially, the results were very satisfactory, but on the other hand any expansion was curtailed.

GE in America was also active in 1896 in setting up the Incandescent Lamp Manufacturers Association, which consisted of GE and six other manufacturers, previously bitterly competing companies. Indeed, after its formation, another ten joined, and the Association's objectives were to fix lamp prices, allocate business and customers to each member. Although Westinghouse was not a member of the Association, there were similar agreements between them and the members. With GE having been the main instigator of the Association, it naturally favoured the larger manufacturers. leaving the smaller ones in a much weaker position, since they did not have access to many of the lesser patents relating to lamp production. This made their lamps generally more expensive and often poorer quality. In 1901, most of these companies merged under one holding company, the National Electric Lamp Company; again, GE cleverly took the major shareholding in National Electric, thereby reducing their ability to compete. To the outside world it appeared that National, as a co-operative of producers, was competing with GE; indeed, credence was given to this when, in 1904, GE brought lawsuits against National for infringement of its patents. These were later withdrawn, but left the public with the impression that GE had nothing to do with National.

GE had an option to purchase the balance of the shareholding in 1910, which it exercised. This was very short-lived as, in 1911, the American Department of Justice brought an anti-trust case against GE, Westinghouse, National, Corning Glass and 31 other companies. The court found that these companies 'are and have been engaged in unlawful agreements and combinations in restraint of Trade'. It recommended that National be dissolved and the assets acquired directly by GE; but this made little difference, as they already had approximately 80% of the market. The court also forbade the supposed price fixing and the forcing of price maintenance on distributors. All this had little effect on the market, as restrictive practices based upon their patents continued; the dominance of GE in the American market was to continue indefinitely.

In the UK, with Swan having lost most of his legal battles and the Ediswan brand dominating the market, other British manufacturers spent little on development initiatives and soon fell behind their Continental and American competitors, as indeed had all British electrical industries. It had been stated:

> *'That the legal obstacle to electrical expansion was removed in 1888. The obstacles which remained, and which largely persisted from 1897 to 1912, were apathy, limited ability and lack of specialisation. The British made no technical*

> *contribution to the development of metal-filament lamps. There was not a single lamp research laboratory in Great Britain during all that time and all-important innovations were imported from Germany, Austria and the USA.'*[1]

Hence, the British industry, influenced by its American partners, was slow to adopt the new metal filaments; in 1905 they attempted to protect themselves from Continental producers by establishing the British Carbon Lamp Association, consisting of:

British Thompson-Houston (BTH)

Cryselco

Edison and Swan United Electric (Ediswan)

GEC

Pope's Electric Lamp

Siemens Brothers

Stern Electric Lamp

Sales under this organisation were on a contract basis between individual manufacturers and individual distributors. The manufacturers maintained minimum prices, up to the wholesaler-to-retailer stage but not beyond. This association tried to copy the Continental and American pricing rings but the legal position was more difficult in Britain than in Germany or America. The three main manufacturers up to the beginning of the First World War were British Thompson-Houston (BTH) a subsidiary of GE, the General Electric Company Ltd. (GEC), formed out of Robertson Electric Lamps, the General Electric Apparatus Company, and Ediswan. Hence the British industry was strongly influenced by foreign companies. While they had some patent rights and cross-licensing agreements, with little development activity between the British manufacturers, it left them more vulnerable to foreign competition.

The introduction of the metal tungsten filaments on the Continent in 1907 gave these manufacturers a much better and more efficient product. As explained earlier, Hirst of GEC was the first British company to negotiate an agreement to manufacture metal (tungsten) filament lamps under the Osram brand. Production commenced in 1909, giving GEC a stronger position. The value of the British Carbon Lamp Association began to be undermined by this development. In 1912 this led to BTH, GEC and Siemens making an agreement, to avoid litigation amongst them, to pool their metal filament lamp patents and jointly issue licences to some other British lamp manufacturers. Later in 1912 a new association was formed, the Tungsten Lamp Association, which was formed with manufacturers using the tungsten metal filaments.

This Tungsten Lamp Association comprised of:

BTH
GEC
Siemens Brothers
Edison & Swan United Electric (joined later)

This Association was founded primarily on patent pooling, with the declared objectives being to promote and protect the interests of manufacturers, to assist in development and research for the lamp improvements, and to make price agreements with wholesalers and retailers.[2]

The key processes in the new metal filaments were developed and exploited by German companies, and because of their much-increased efficiency their sales and export rose strongly, putting them ahead of any other European country. Their production in 1908-1910 almost rose to that of the American production, being a third of world production. However, their patent protection was weak, increasing competition; and falling standards endangered the profits of the main producers. Hence, in 1911, this led to the three largest producers, AEG, Siemens & Halske and Auergesellschaft pooling their metal filament patents and coming to an agreement on price maintenance through the *Drahtkonzern*, (Filament Trust). This dominated the European lamp industry until the outbreak of war in 1914, the decisive factor in the breaking-up of most if not all the international rings. As described earlier, these three companies amalgamated in 1909 to form Osram.

During the 1914-18 war the production and distribution of lamps in the United Kingdom was regulated to a considerable extent and largely protected against imports. During the war there were periods of shortages, and the government Sub-Committee investigating the industry in 1920 did satisfy themselves that in these periods of shortage, the Association did have a stabilising influence on electric lamp prices and preventing prices from rising. However, by the end of the war it was clear that several companies had concentrated on carrying out research on metal filament lamps, and several new patents been taken out on these developments. It was considered that several of these patents overlapped and costly litigation was continually threatened. In 1917, BTH claimed a monopoly to manufacture drawn tungsten filament wire under their patent rights, which was one of the basic patent rights of the Tungsten Lamp Association. A non-Association company, Duram Ltd., challenged these patent rights in the High Court of Justice, the Court of Appeal and in the House of Lords and they won their case. Therefore from 1917 in the UK, patents no longer restricted the manufacture of drawn tungsten wire. Indeed, this led to modifications in the law contained in the Patents and Designs Act in 1919, designed to provide a remedy for abuses of patent monopolies.

Electric Lamp Manufacturers Association of Great Britain Ltd

Deciding that co-operation was preferable to conflict, in 1919 the Carbon Lamp and Tungsten Lamp Associations merged to form the *Electric Lamp Manufacturers Association of Great Britain Ltd.* controlling 95% of the market. The objectives of this Association were: 'To formulate, regulate and secure uniformity of practice in the manufacture, sale and purchase of electric lamps in the United Kingdom, in such a manner as to benefit both the trade and the public by the adoption of standard conditions of sale and of product'. Under these provisions the Association in practice fixed common retail prices and trade terms based on annual quantities purchased, and maintained a system of exclusive agreements. This exclusivity and price maintenance was supported by a register called the 'List Prices Net Register', known in the trade as the 'Black List'. These were people/organisations that were only to be supplied at full retail prices. The founder members were:

BTH
Edison Swan Electric (name changed in 1916)
Foster Engineering
GEC
Stern Electric Lamp
Siemens Brothers
Z Electric Lamp Manufacturing

British Westinghouse (name changed to Metropolitan Vickers Electrical [Metrovick] in 1918)
Dick, Kerr (Britannia Lamp & Accessories)
Popes Electric Lamp

In 1917 British Westinghouse purchased Brimsdown Lamp Works from the Custodian of Enemy Property. In the following year it changed the name to Metropolitan Vickers Electrical Co. Ltd., and became known as Metrovick.

More significantly, with the market expanding, GEC, along with other members of the 'Ring', were making good profits, so it was no surprise that Hugo Hirst worked hard during the summer of 1916 to acquire the other half of Osram from Auergesellschaft from the Custodian of Enemy properties. In August 1916 he managed to complete the deal, making the Osram brand in the UK and the Colonies wholly owned by GEC.

As outlined in the earlier chapter on 'Industry Consolidation', a financier, C. B. Crisp, purchased the Siemens & Halske share of Siemens Brothers from the Custodian of Enemy properties in 1917. Siemens Brothers had been established in the UK since 1858, and through family links, had maintained a close relationship with the German company of Siemens & Halske. This created a board of directors with little or no knowledge of the lamp industry.

A number of non-Association companies meant that the British lamp industry was divided into two divisions with the smaller companies being excluded from the benefits of cross-licensing available to members of the Association. The main non-Association companies were:

Cryselco Lamp Co. Ltd.

Crowther & Osborn Ltd.
Imperial United Lamp Co. Ltd.
Corona Lamp Works Ltd.
Harlsden Lamp Co. (Stella Lamp Co.)
Maxim Lamp Works Ltd.
Notable Lamp Co Ltd.

In 1920, only one year after its formation, the government Sub-Committee on Trusts[3] found that the Association controlled between 90-95% of the industry output. It also found that while the Association would not necessarily pursue a policy of inflated prices and inordinate profits as a result of its position, there would be no effective check in the shape of competition should it decide to do so. This Sub-Committee found 'immoderate profits' had been made on the manufacture of lamps before 1914; discounts allowed to distributors ought to be reduced. It implied that non-Association manufacturers, while still making satisfactory profit, were able to sell at lower prices than Association members were. Since the Association retail prices were universally adopted, the effect of these lower prices was reflected in the terms to the wholesaler/retailer rather than to the public.

The Sub-Committee further drew attention to the close connections between patent rights and the formation of the ELMA of Great Britain Ltd. and the granting of licences by the patent-owning members to other members against payment of royalties. This continued acceptance by members with licensed patents, held to be invalid by the High Court which imposed a restriction on the output of lamps manufactured for the home market under licence. This not only applied to the manufacture of lamps but also to components. Hence, in spite of the judgment given in the Duram case mentioned earlier, the ELMA members had an agreement not to dispute the patents of the licensors and to buy their drawn wire only from an ELMA company, although the patents were declared invalid. This meant that the home market was virtually closed to Duram. The Sub-Committee regarded these methods of limiting expansion as more harmful than a simple percentage quota system, though they were aware that there were objections to this method also. It was significant that they drew attention to the danger that the interests of the British industry would be subordinate to American interests and that an international combination might 'control supplies and dominate prices over a considerable part of the world'. With this in mind it inevitably recommended that 'the operations of an Association which so effectively controls an important industry should be subjected to public supervision and control'. There is no evidence that this recommendation significantly changed the objectives of this Association. In 1921, BTH and GEC had concluded an agreement providing for non-exclusive cross-licensing in Britain of all patents owned and controlled by each party and exchange of information. GEC had the benefit of the GE information and patents in Britain through BTH. Though excluded from the American market GEC would pay royalties on all sales in non-exclusive territories. Thus from an early date, GEC and BTH had the benefit of the inventions of the principal German and American manufacturers and a guarantee from competition from them in their own home market. In return, the British manufacturers undertook to keep out of the much larger exclusive market of its foreign partner and, for use of their inventions, paid royalty to their partner.

Similar individual cross-licensing arrangements existed between lamp manufacturers in continental Europe; indeed,many went back to the early 1900s. While these arrangements often included joint selling agreements with fixed prices, they continually broke down because of outside competition undercutting them. The 1914-18 war significantly changed the market and new productive capacity had been developed, particularly in neutral countries; most notably, Philips in Holland were able to create a world-wide market. New frontiers were soon being protected by high tariff walls, which frequently separated manufacturers from their normal markets. Europe had created considerable surplus productive capacity, so that dumping at prices representing cost or less became a common occurrence.

In 1922, Cryselco and British Electric Lamps (BELL) joined; however, this was followed by Foster Engineering, Stern Electric Lamp and Z Electric Lamp Manufacturing ceasing to manufacture and they became licensees. This enabled them to have 'manufacturers' discounts to sell to all classes of customers, even ELMA of Great Britain Ltd., who were distributors. This privileged purchasing was never to be withdrawn.

Philips was becoming a much stronger importer of gas-filled lamps, known as Half-Watt lamps; in 1919 the Board of Trade granted licences for the importation of 1.25 million lamps to GEC, Siemens Bros. and BTH. These lamps were bought for 3s. (15p) and retailed for 12s. 6d. (62.5p), compared with 3s. 6d. (17.5p) for normal vacuum lamps. This was subject to criticisms of the government sub-committee in 1920, who recommended that these margins were excessive and that these lamps should retail at 8s. (24p).

Philips later imported large quantities of lamps under the brand 'Condor'. After lengthy negotiations, these imports ceased; in 1925

Philips finally joined ELMA of Great Britain Ltd., becoming an established brand, although initially the lamps continued to be imported from Holland.

Although Crompton Parkinson entered the lamp market with the purchase of Nox Electric Lamp Co. in 1929 and moved to a new factory in Guiseley in 1932, they did not join ELMA until 1937. Aurora Lamps joined in 1936.

In the late 1920s, Tungsram Electric Lamp Works from Hungary also joined ELMA of Great Britain Ltd. and after acquiring Orion Lamps they marketed lamps under both the Tungsram and Orion brands. They continued to gain market share with lamps made in Austria; they had factories not only in their home country of Hungary but also in Czechoslovakia, Italy and Poland. With the introduction of the general tariff in 1932, it was found impossible to continue to import lamps and under an ELMA of Great Britain Ltd. agreement, they withdrew from the British market and received compensation for so doing.

The gas-filled lamp patent was to expire in 1929, but upon application was renewed for two years. The independent manufacturers were therefore granted a licence provided they sold lamps at the ELMA of Great Britain Ltd. prices, with the exception of Nox Electric Lamp Co. who were excluded from making gas-filled lamps.

The home market was partly protected against competitive imports until 1931 by the gas-filled patent. In other countries, war-time expanded the lamp industry, and in the years following 1918, this led to fierce competition, particularly in Europe, where much business was carried out at production cost prices. Indeed, in Europe in 1921, the International Glühlampen Preisvereinigung (a price fixing cartel) was formed in Germany. International General Electric (IGE), formed in 1919 to represent GE's interests outside America had been represented in this, when the cartel fell apart in 1924. The first executive president appointed to manage IGE was a very ambitious and powerful man with a very high intellect: Gerard Swope. It was clear to Swope that other pricing production/strategies must be pursued if he was to protect not only the USA, but also all the overseas territories in which GE was represented. Having travelled widely throughout the world and being widely respected, he was appointed as President of GE. The role of guiding the lighting interests of IGE fell to J.M. Woodward, who was the Director of IGE Europe.

In 1924, the leading lamp manufacturers in the world, led by J.M.Woodward of IGE, but almost certainly strongly backed by Gerard Swope, the president of GE, negotiated what was eventually known as the *Phoebus Agreement*, which is covered later. This was only possible because the lamp industry was substantially different from other commodity industries in that it was based on patents, licences and cross-licences. By concentrating control of patents and extending licensing agreements, companies could exert considerable influence on the control production and sale of incandescent lamps.

Electric Lamp Manufacturers Association ELMA

In 1933, the members of ELMA of Great Britain Ltd. formed a new unincorporated association, the Electric Lamp Manufacturers Association, ELMA, and registered it as a trade union with members being confined to manufacturers. This new Association took over and maintained the system of prices and the exclusive agreements formerly operated by ELMA of Great Britain Ltd., leaving the old Association in existence to hold property leases and to fulfil certain educational functions. The Association increasingly concentrated its emphasis on standards and quality of production. It also entered into arrangements with distributors designed to encourage them to co-operate in its arrangements for regulating distribution. This extended to seeking agreements with distributors on prices of imported lamps, particularly Japanese, which were being distributed at low prices through retail shops, most notably the Woolworth chain. A typical retail price for general service lamps (GLS), in 1935 in Woolworth's shops would be sixpence to a shilling (2.5p-5p); this would be compared to an ELMA retail price of 1s 7d to 1s 9d (8p-9p). With such a price differential, the promotion of quality could hardly be expected to deter the increasing market share of the cheaper lamps. In response to the growing complaints about their high prices, the ELMA members arranged in 1935 to produce limited quantities of a type 'B' lamp, designed to have a lower quality than the normal 'A' lamps and to retail for a shilling (5p). By this time, Woolworth had become the key retailer of lamps; and a non-ELMA manufacturer, Britannia Lamps, won the contract to supply them. Britannia was soon acquired by Ismay Industries and by 1938 they became the largest independent manufacturer in Britain. It was at this point that ELMA members stepped in and purchased the lamp interests of Ismay Industries (consisting of Britannia, Ismay and Gnome).

The independent manufacturers were generally small concerns, producing a more limited range of lamps, although there were some notable exceptions. One such company, Crompton, built up a successful business in lamps at ELMA prices and in lamps retailing at 1s. 0d. but they became a member of ELMA in 1937 as part of the settlement of litigation for infringement of patents. By the start of the 1939-45 War, there were two distinctly priced markets for lamps, the premium brand, type 'A' and second brand, type 'B', often supplied by the ELMA members. There were still a number of independent manufacturers remaining outside ELMA. These included Jules Thorn's Electric Lamp Service Company, which merged with Atlas Lamps and, with several smaller companies Luxram, Maxim, Insular, Evenlite, etc., remained outside ELMA.

Another significant aspect of the electric lamp industry was the development of its own machinery/manufacturing equipment and component manufacture. This played an important role in the growth of the larger companies. Manufacturing techniques in the early development of the industry were closely guarded and linked to patented developments and these required dedicated components. The early pioneers and entrepreneurs preferred to be in control of the whole product. The consequence of this was particularly prevalent in the 1914–18 and 1939–45 wars when the members of ELMA, like their international competitors, developed the production of lamp components and especially the machine-made glass bulbs and tubing, caps, tungsten and molybdenum wire. In the UK, two members, GEC and BTH, owned the only factories producing machine-made bulbs and one of the two cap factories; the other cap factory, was set up during the 1939-45 war and was owned by the largest independent manufacturer Thorn. In general, the independent manufacturers had to rely on the supply of components from outside sources, mainly from imports or during the war, from ELMA members.

The lamp industry was regarded critical enough during the war to be placed under the control of the Ministry of Aircraft Production; later, the Ministry of Supply assumed control of production, allocating materials and available labour between all manufacturers and setting production targets. This control was removed in 1945, but a system of voluntary allocation of glass bulbs was reintroduced in 1947. During this period, filament lamp prices were subject to the Prices of Goods Act 1939, in which the net cash profit on the sale of a given type of lamp was restricted to the amount of profit obtained in August 1939. From 1948 to July 1949, there was a Maximum Prices Order, which imposed a standstill on certain lamp prices, in particular general lighting service (GLS) lamps, but not motor-car lamps. Discharge lamps had not been subject to any price control, and in July 1949 all statutory price control of filament lamps ceased.

The General Patent and Business Development Agreement (Generally known as the Phoebus Agreement)

In the post-1914-18 war period, the German companies wanted to re-establish themselves; significant increase in capacity had been created in many of the major companies, and the inevitable result was fierce price competition largely due to over-capacity. In 1921, the newly formed Osram GmbH Kommanditgesellschaft, a limited partnership between AEG and Siemens & Halske each having 40% and 20% held by Leopold Koppel the proprietor of the Auer company, decided to take the lead in re-forming the Continental ring that was first set up in 1903. This was to be named the International Glow-Lamp Price Organisation. At the same time, Osram GmbH and GEC agreed to exchange patents. Osram GmbH and GEC made similar agreements with GE. In the same period GE

France broadened its influence and the three leading electrical companies merged their lamp interests to form Compagnie des Lampes, becoming the dominant French company. The combination was similar to that of Osram GmbH some two years earlier.

This cartel got off to a good start, but it soon began to falter due to the aggressive competition between Osram GmbH and Philips, which finally resulted in a complete breakdown in 1924. This was due to Osram GmbH wanting to regain markets lost during the war and Philips naturally being unprepared to relinquish them. Since the early 1920s, GE had also been very active in strengthening its position in America, and always fearing the influence that uncontrollable foreign lamp markets would have on its own domestic market. In the post-war periods, GE sought to create an international organisation, to regulate prices, to set output, and divide markets. International General Electric (IGE), formed in 1919, represented GE's interests outside America and a very influential man, J.M.Woodward, headed the IGE lighting activity.

Throughout 1924, IGE had been actively negotiating with Osram, Philips and Vereingte Glühlampen & Elektrizitäts A.G of Hungary. Indeed it was Woodward who became the chief negotiator for IGE, and in September 1924 wrote: 'It may require considerable manoeuvring to get a formula adopted which will be accepted by the Germans and Philips.' Eventually these negotiations led to the basis for the Phoebus Agreement involving most leading lamp manufacturers throughout the developed world. Many of these manufacturers were brought in at a late stage, including the British Group, when most of the arrangements were already made.

With competition being extremely fierce and the danger of it increasing, an agreement was eventually signed in December 1924. This agreement, driven by J.M.Woodward, was reached with all the major manufacturers of the world to create a pricing and quota agreement called the *'General Patent and Business Development Agreement'* and it was incorporated in Switzerland.

The day before the agreement was signed, Woodward sent the following letter describing the negotiations to the president of GE:

> The process has been one of education, and as that education progressed I have taken advantage of the entire freedom of action which you left me, to develop the affair into a living partnership. For a long while it appeared as if the only thing, which it would be possible to achieve, would be a naked agreement between the parties as to quota. Especially this seemed to be true at the time when the personal relations became strained to the point of passing very strong epithets back and forth. Gradually the European owners and managers came to like the idea of a permanent relationship...and took advantage of every opportunity to advance this idea... Also they have become impressed, some more than others, with the fact that they must co-operate in what we call the Business Development program, and have become educated not so much in the mortality of this co-operation as in the necessity for it... I must say without being misunderstood that the European parties are entirely amateurish, for reasons, which arise from their historic division of interest. They realise this, and that realisation coupled with their still active mistrust of each other has caused them to accept our leadership and to

> invite its continuance. This situation gives us an opportunity, which I do not think we can neglect. The entire group can, in my opinion, with careful and intelligent handling on our part, be ultimately organised into at least an inseparable coalition and probably into something better still, within the next nine or ten years at our disposal. This result does not depend upon the chance of actual mergers between the component parts since it can be reached otherwise by taking hold of the enterprise in its beginning and never letting it get away from us through slack or indifferent manipulation.'[4]

The General Business Development Agreement was set up as a Swiss registered company under the name of Phoebus S.A. Compagnie Industrielle pour la Développement de l'Eclairage, registered in Geneva; this became known as the *Phoebus Agreement*. All participating manufacturers contributed to the shareholding of Phoebus in proportion to their lamp sales of the total world sales. This was only possible because the lamp industry was substantially different from other commodity industries in that it was based on patents, licences and cross-licences. By concentrating control of patents and extending licensing agreements, companies could exert considerable influence over the control of production and sale of incandescent lamps. This Agreement was legally set out in some 22 Articles of contract; the following is a summary of this very detailed Agreement.

Neither GE nor IGE were formally a party to the Phoebus Agreement. In fact GE, including IGE, was the only major world lamp company not to sign this agreement. However, years later, in the late 1940s, the reported opinion of the judge of a District Court in a Civil Action (no. 1364) of the United States of America

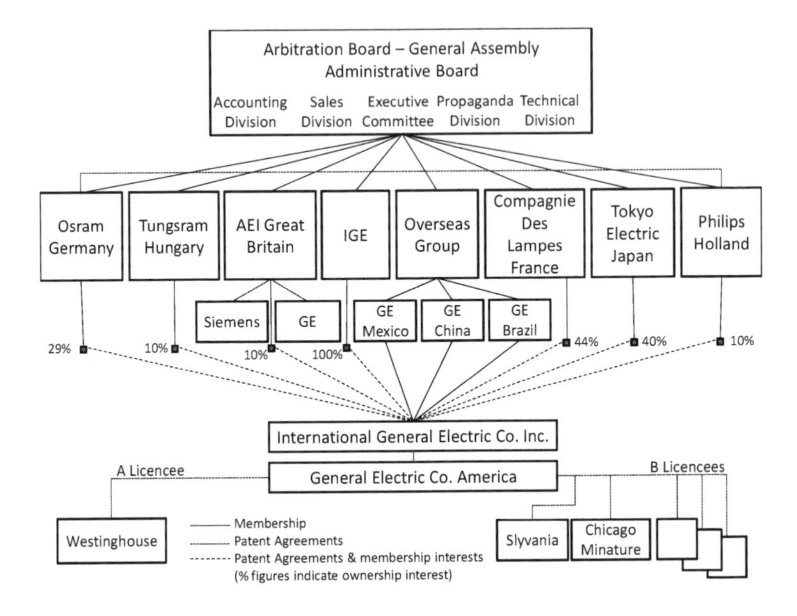

Figure 36: Phoebus structure

versus General Electric Company of America, was that GE was the hub of 'a gigantic world cartel'.[5] He regarded IGE to have played a prominent part in the negotiations, the complementary agreements and the actual adherence to that agreement. After the conclusion of the Phoebus Agreement, IGE made new agreements with most of the principle parties for the exchange of patents and technical information, recognising the United States and Canada as the exclusive territory of IGE. Exclusive rights were granted to the other parties in their own territories, so far as IGE was concerned this included the mutual grant of non-exclusive rights in common markets.[6]

One such agreement was that between IGE and AEI and its subsidiaries, BTH, Ediswan and Metrovick. Made in 1930, this agreement superseded the one made between GE and BTH in 1896 and 1905, renewed in 1919 with IGE. In effect this contained five significant clauses:[7]

- IGE may not manufacture or sell in the UK market.
- AEI and its subsidiaries may not manufacture or sell in North America.
- The parties license one another exclusively in their respective exclusive territories and non-exclusively elsewhere, except in so far as IGE has already granted exclusive rights to other manufacturers.
- There is complete exchange of manufacturing information.
- AEI and its subsidiaries pay IGE a fee based on factory cost of the lamps they make in the UK.

The purpose and intent of the Phoebus Agreement was to secure the co-operation of all participating companies in:[8]

1. Providing a more advantageous utilisation of their manufacturing facilities in producing lamps.
2. Securing and maintaining a uniformly high quality of products.
3. Arriving at more economical arrangements for product distribution.
4. Promoting more efficient of electric illumination and the increased use of light to the advantage of the consumer.

All parties were to continue their business on independent lines with the provision for each company to secure its proper share of the increased business from the co-operative effort.

An important aspect was the adoption of the recommendations of the Incandescent Electric Lamp Manufacturers' Technical Association made at the meeting held in Paris, April 1924, with respect to standardisation, lamp efficiency and other related matters. This aspect of agreements on international standardisation remained with the most industries up to the present day.

A detailed procedure for exchange of inventions and experience, covering licensing of patents to all participating companies was established. Included in this procedure was a method of handling disputes or disagreements via a Tribunal of Arbitration.

A method was established for determining the sales and relative market position of all the companies by taking the total sales in the territories covered by the Agreement in one of the proceeding years 1922 or 1923. Again, sales were carefully defined so as to exclude all intercompany sales and only to include genuine third-party sales. For new members applying to join the General Agreement from 1930, data was taken based on their average sales of the previous three years. A special formula was agreed in which each lamp/class of lamp was given an index covering all incandescent lamps from miniature to 10,000w, including projection, photographic, gas-filled, vacuum and filament lamps for special purposes. Therefore total sales of all parties in the territories governed by this Agreement were fixed. With the percentage sales of each party being established, not only in total but also by country, this was to remain unchanged for the period of the Agreement, any change to be made by unanimous agreement. This data established a General and a Local Participating Percentage; a committee appointed by the General

Board would monitor this data and take into consideration the prices realised for the lamps.

Globally, three geographical territories were identified:

- Home Countries, in which the Member Company had its most important manufacturing facilities or sales.
- British Overseas territories
- Common territories

Officially America and Canada were excluded from any allotments based on sales and hence, by implication, can be regarded as a fourth territory. It is also worth noting that while all the world's leading lamp manufacturers signed the Phoebus Agreement, there was one exception, GE of America who did not sign; although J.M. Woodward of IGE, being the prime negotiator, did sign the outline agreement subject to the approval of his principals. As GE was predominant in America/Canada and its policies were thoroughly intertwined with those of Phoebus, it made no difference. Indeed the interchange of patents was the feature of a separate agreement between IGE and the members of Phoebus for which it collected five million dollars in royalties and service charges between 1919 and 1941.

Home Countries were defined as the following, for which no additional groups could be formed including these countries:

Austria
Belgium *(incl. Luxemburg from July 1932)*
Brazil
China *(incl. Manchuria, Mongolia, Tibet and Chinese Turkestan)*
France *(incl. The French Colonies and Protectorates and Monaco)*

Germany
Great Britain *(incl. Ireland, Channel Islands and Isle of Man)*
Holland
Hungary
Italy *(incl. The Italian Colonies and Protectorates)*
Japan
Spain

Sales to the United States and Canada were to be excluded, being defined as GE's home territory.

British overseas territory consisted of the British Empire excluding Britain and Canada. Sales within this territory were the responsibility of The British Group, plus Osram, Philips and Tungsram.

The British Group consisted of:

The British Thompson-Houston Company Ltd. (BTH)
Cryselco Ltd
Edison Swan Electric Company
General Electric Company Ltd. (GEC)
Metropolitan-Vickers Electrical Co. Ltd.
Siemens & English Electric Lamp Co. Ltd.

Common territory was the rest of the world, excluding the home territory, United States and Canada, and British overseas territory.

Without doubt a major disadvantage of this Agreement was the considerable effort that went into maintaining the relative position of all participating companies. Of all the Articles making up this very formal legal agreement, this will have taken the most time and discussion to establish and maintain. Each company participating in a country or established group of countries was entitled to make unit sales corresponding to its Local Participating Percentage (LPP). All excess sales were regarded as growth due to co-operative effort and therefore the profit was divided between the companies in deficit in that territory. Detailed formulas were established for measuring average realised prices, sales, and hence penalties to be paid, should the LPP be exceeded. Hence, such payments on excess sales would be divided among the parties participating in such territory whose sales in that territory had been less than the number of units corresponding to their Local Participating Percentage in proportion to their deficits.

This guaranteed a secure domestic position and a portion of sales in each area. Phoebus placed no limit on total sales; however, each member had to stay within its designated quotas for each distribution area. If sales exceeded the allotted quota, the profits on excess were penalised. The penalties were distributed to those companies that did not meet their quotas in proportion to the difference between realised and the quota. Through this system, Phoebus advised, but avoided the need to fix prices in each individual market; it transferred this responsibility of fixing prices to the National Associations composed of each firm in a given market area.

It was a fundamental principle that each company should continue to enjoy the volume of business corresponding to its General Participating Percentage. It was agreed that whenever the average world prices showed a price ratio at variance with the existing ratios set out in a schedule in the original agreement, then this schedule would be updated to establish new price ratios. This would be examined in periods of not less than six months.

In France, the businesses of Compagnie des Lampes (with its

licensees) and of Philips would be considered a bloc. Similarly considered were the Societa Edison Clerici for Italy, the Tokyo Electric Company for Japan, the Osram Company for Germany, Compagnie des Lampes for Spain, the Philips Company for Holland and Belgium, each having the right to make arrangements with its licensees in the Home Country. The result was that each company in its Home Country could form a bloc should it so wish.

Similarly, the businesses of the manufacturing companies of the ELMA, referred to as the British Group, were considered as a bloc, so far as Great Britain is concerned. They joined to form a bloc with their agreed licensees. Only licensees who agreed to conform to the terms of the Agreement could be included in a bloc.

Finally, the businesses of the companies forming the Overseas Group were also considered a bloc for all purposes.

This Article, more than any other aspect, was to absorb a tremendous amount of unproductive time. It contributed to complacency in development of all aspects of the business. In a study carried out in the USA in 1998, Patrick Gaughen on 'Structural Inefficiency in the early Twentieth Century' concluded:

> *Economic inefficiency in the early twentieth century was primarily oligopolistic in nature. Both the aluminium (the other industry) and incandescent lamp industries were characterised by monopoly or price leadership oligopoly. These industries were classified internationally as colluding oligopolies. These forms of market structure were economically inefficient because they restricted output and inflated price. They were socially inefficient because they failed to pass on the benefits of technological improvements to the consumer and the public.*

Additionally, there was a provision to make financial adjustments necessitated by actions of a party that was prejudicial to the interests of others. This could apply to a participating company disadvantaged when actively furthering the common interest when other parties do not fairly or fully co-operate by following the policies and procedures agreed upon. It was agreed that if one of the parties should in any way gain or try to gain advantages over other parties by actions not in accordance with the Agreement, then the other party/parties would be adequately compensated for any damage from these actions.

As a surety for the fulfilment of the obligations incurred by the companies signing this Agreement, and for their share of the liability in the cost of putting it into effect and eventually to provide for the adjustments and payments, each company agreed to deposit negotiable securities or instruments, 10% of which must be in cash or equivalent. The initial amount for each signatory party was $4000 plus a further $4000 for sales of each complete million units above the first million. This was for any party that might be in default of its liabilities. In addition, there was a second series of sureties, of which 10% must be in cash or equivalent, and increased by $4000 for sales of each million units above the first million. This second series of sureties was held to compensate parties unfairly penalised by another who infringed the Agreement. All sureties would be returned at the termination of the Agreement.

There was even an Article enabling the General Board to hand out fines against parties who persistently neglect or omit to submit their reports or otherwise hinder the daily working of the joint business of the parties.

Great care had been taken in setting up the organisation and the General Board. The main body with supreme authority was the General Meeting, which met once a year and consisted of all signatories to the Agreement (Appendix 1). It delegated its decisions to a General Board, which also had responsibility for the administration of the Agreement. The General Board was made up of representatives from the following companies:

Osram Group	3 Representatives
Philips Group	3 "
Overseas Group (IGE)	2 "
British Group	1 Representative
Compagnie des Lampes Group (France)	1 "
Kremenezky Group	1 "
Societa Edison Clerici Group (Italy)	1 "
Spanish Group	1 "
Swedish, Swiss and German Groups	1 "
Tokyo Electric Co. Group (Japan)	1 "
Vereinigte Group (Hungary)	1 "

The General Board elected a Chairman and Vice-Chairman each year, and appointed a General Manager to be responsible for the daily administration of the affairs of the organisation; this General Manager became the effective executive head and an ex-officio member of the General Board. In addition, the Board established two further committees, an Executive Committee and a Development Committee. The Development Committee was made up of six technical and six commercial members, and its function was to decide on the general policy to be followed in matters of lighting development, technical development and standardisation for information at Local Meetings. These Local Meetings were established in Home Countries to watch over special interests in that country related to the Agreement. Representation and voting power was according to the Local Participating Percentage.

As an example, the British Group managed Local Meetings in Great Britain, and all decisions relating to business consisting of:

- Prices, sales, rules and conditions.
- By-laws, rules and regulations for the conduct of Local Meetings.
- Method of voting in Local Meetings within the rules of the General Board.
- Implementing the method of maintaining the relative position of the Group (Article 6 of the Agreement). This implied that maintaining the minimum sales price would be adopted by them in line with Article 6 and that they establish the price for tenders to the Main Departments of the British Government. Also the agreement that if the percentage of the minimum sales price representing the entire profit be adopted by them and not be acceptable to the General Board, then this percentage shall be ascertained by the auditors.

There was a separate codicil relative for an agreement, made in July 1922 for five years, between Philips and Edison Swan Electric Co. Ltd. This laid down that all methods, information, models, plans, data, drawings, experience, advice or assistance given by Philips to Edison Swan for their own use was to be kept confidential by them at all times. A similar codicil existed concerning the transfer of technical information from Philips to

Cryselco. These would be typical of the rules for Local Meetings to manage in Home Countries.

Other Articles of the Agreement covered appointment of accountants and auditors, submission of sales/financial statements, acquisition/sale of businesses, withdrawal from the Agreement and settlement of disputes. Each signatory had to take the necessary legal advice to ensure they and their Company were able to carry out the requirements of the Agreement.

There were in all 22 Articles of Agreement, and the final Article set the legal structure for the implementation and operation of the Agreement. The parties to the agreement had to bind themselves to observe the terms and conditions in the spirit of mutual co-operation, fully realising that in specific cases individual sacrifice may be demanded in return for benefits which would ultimately accrue for their common good.

In this final Article, the parties to the Agreement appointed a Limited Company (société anonyme) Phoebus S.A. Compagnie Industrielle pour le Développement de l'Eclairage, having its principle office in Geneva, Switzerland. This became the 'Bearer of their Rights' (Porteur des Droits) in so far as it had the right to enforce on their behalf and in the name of Phoebus their common and individual rights under the Agreement. The parties to the General Agreement paid the share capital of the Phoebus S.A. in accordance with their General Participating Percentage. Finally, it was also stated that in the eventuality of liquidation of Phoebus S.A. or termination of the General Agreement, the assets or liabilities would be divided according to the General Participating Percentage.

In 1928, a Swedish company drew up plans for a lamp factory.

Despite threats of market muscle and patent infringement suits from Phoebus, the Swedes went ahead with the factory. A co-operative union, Luma, sponsored by smaller companies in Denmark, Sweden and Norway, began lamp production in 1931. It was able to produce and sell lamps at price levels substantially below those of the Phoebus companies and to make a profit.

In the United Kingdom, the British Group with its dominant position, had quota arrangements, including control of prices and terms of sale. The participation of Crompton in the Phoebus Agreement in 1937 did not substantially alter this position, although it amounted to recognition of Compton's right to retain the share of the trade it had previously built in opposition to the Phoebus parties. GE was prevented from competing in the UK by individual agreements. The members of the British Group also had the benefit of the patents and manufacturing know-how of GE and Osram, but in exchange not only imparted their own knowledge and inventions to those companies but also paid them fees and royalties. It was not until 1938 that they made some limited agreements with Philips providing for cross-licensing without royalties on either side.

While pricing responsibility was given to each individual market, they nevertheless became inflated due to concentrated control over the market. Therefore it was not surprising that in the latter part of the agreement prices and not production became paramount for profit in the cartel arrangement. The following is a quote from a Westinghouse memo in February 1937:

> *In all countries...where we do business in the common territory we invariably have a larger percentage of the market than our percentage in the Common Territory, consequently, if the*

> *growth of business is greater in those countries than in the Common Territory as a whole, the position becomes more embarrassing for us as regards exceeding our permissible sales. Unless one can purchase units at a very reasonable figure – and it is not always easy to do this – it becomes most unprofitable to exceed one's permissible sales. The goal therefore to be aimed at is to make as much money as possible out of those units we are permitted to sell. In other words, it is much more advantageous for us to make a profit of 5 cents per unit on 4M units than 2.5 cents on 8M units.*

Hence artificial inflation of prices not only became evident when examining relationships between the firms within the Phoebus cartel, but also when examining the competition that arose outside Phoebus. The electric lamp was in many ways a product fitted to a producer's cartel. It was a necessity for which there was no alternative, and the price was inelastic. Thus any cut in price, which was a relatively insignificant part of the average domestic budget, would not lead to a rise in demand because, as today, the amount of light people used was governed by the price of electricity rather than the cost of the bulb. The exact effect of the lamp ring on prices to the consumer has never been adequately calculated; as such, calculations have to include various hypothetical assumptions.

There is no evidence of any significant dissent with the philosophy or important policies of the Agreement but if there had been, GE, through its shareholding in the various participating members, ensured that any problems or dissension was quickly resolved.

Phoebus was in existence for fifteen years but in its latter years a number of members began to see it as restrictive. It was eventually abandoned in 1939, when the American Justice Department moved to break the GE control of the domestic incandescent market and its participation in Phoebus. Although neither GE nor IGE were formally a party to the Phoebus agreement, it was said in a judgment brought by the Justice Department against GE under the anti-trust laws of the USA, 'Civil Action Judgement No. 1364' that in the judge's opinion, GE was the hub of: 'a gigantic world cartel ', having regard to the prominent part played by IGE in the negotiations, the complementary agreements made by IGE with the various parties to the Phoebus Agreement. With the outbreak of the 1939-45 War the relationships of many remaining companies were broken, although some did continue and other new ones were created.

The electric lamp has been a huge stimulus for the early growth of the electrical industry and was the product on which several huge global companies were formed. The most notable were GE of America, Philips of Holland, and Osram of Germany. While a significant characteristic of the lamp industry has been lamp rings/cartels, the electric lamp has also stamped itself on most of the rest of the electrical industry. These large lamp manufacturers used their products to diversify into the production of many other electrical products and in most cases these have completely dwarfed the lamp side of their businesses. Already before the 1914-18 War, GE and Westinghouse of America, AEG and Siemens of Germany, French Thompson-Houston, British Thompson-Houston and GEC of Britain were producers of a wide range of electrical products, ranging from small consumer products to heavy electrical equipment such as generators. Another characteristic of these companies was their vertical integration, and this took them into the manufacture of the components, such as wire, chemicals, glass and metal components. This is demonstrated in a later chapter, showing how it also influenced that development of the industry and still does so to this day . This either spawned them as suppliers of these components to other industries or gave them the expertise to manufacture other products themselves, based upon these manufacturing technologies.

The 1948 Lamp Agreement

The outbreak of war in 1939 made the Phoebus Agreement ineffective and the agreement between GEC and Osram was terminated. By 1941, the British Group with A.C.Cossor Ltd., Crompton, Philips (Holland), Stella and IGE entered into another agreement called the 'The New Agreement', which substantially maintained the Phoebus arrangements, except in enemy-occupied and certain neutral territories. In 1945, IGE withdrew from this agreement and again the British manufacturers (including Philips (Holland) created a new formal agreement called 'The 1948 Lamp Agreement'. This applied to their trading in the UK and the British Empire excluding Canada. The companies in this agreement and in ELMA (the Electric Lamp Manufacturers Association) were:

British Thomson-Houston Co Ltd (BTH)
Crompton Parkinson Ltd
Cryselco Ltd
Edison Swan Electric Co Ltd
The General Electric Co Ltd (GEC)
Metropolitan Vickers Electrical co Ltd (Metrovick)
N V Philips Gloeilampenfabrieken (Philips)
Siemens Electric Lamp & Supplies Ltd (Siemens)

Stella Lamp Co Ltd

Philips UK, BELL (Wimbledon) and Aurora (Scotland) were ELMA members but they were not part of the 1948 Lamp Agreement;, Philips Holland had signed the Agreement, and although it was not in ELMA, this effectively bound Philips UK to the Agreement. British Luma had been consistently refused membership because they objected to some of the principles and methods by which ELMA operates.

The percentages shown are of ordinary capital.
IGE also held shares of ordinary capital in the following:

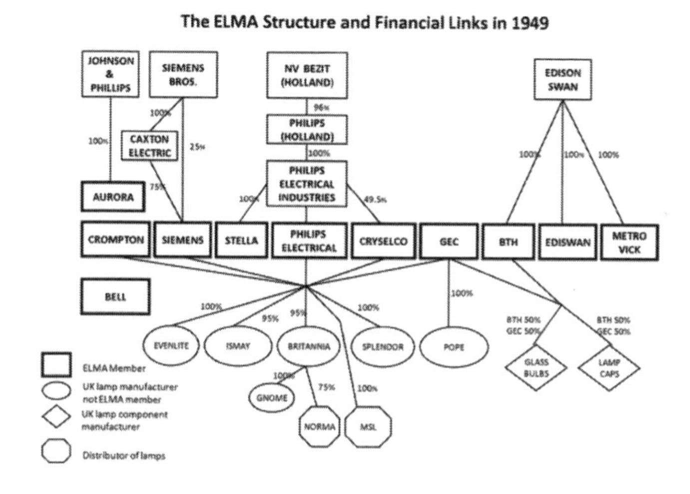

Figure 37: The ELMA Structure 1949[10]

AEI – 27%

NV Bezit (Holland) – 3%

Philips (Holland) – 3.5%

Evenlite, Ismay, Britannia, and Splendor became known as the Controlled Companies (see Chapter 6).

Norma Electric Co. Ltd. only purchased ELMA lamps for Christmas tree lighting sets.
MSL was Splendor's distributing company for motor-car lamps.

Like the Phoebus Agreement, this had been structured on a formal legal basis with 26 Articles, taking much of the structure from Phoebus but obviously tailoring it to suit the new circumstances. The Agreement maintained the quota arrangements for those territories to which it applied but established separate quotas for the UK market. An Executive Committee was set up, also sub-committees for statistics, legal matters, standardisation and technical development. Furthermore, an 'Administrative Office' was established, which set up the Electric Lamps Statistical Office Ltd. to assemble statistics for calculating quotas, penalties and compensation. The Agreement also precluded the British manufacturers from exporting to North America but they were protected from competition from GE in the UK. This however was short-lived as, following the United States anti-trust action against GE and IGE, the UK market was open to GE and the North American market to the British manufacturers.

The scope of the 1948 Lamp Agreement covered all electric lamps for illuminating, heating or medical purposes operated by:

- Incandescence of a refractory filament
- Cathode incandescence
- Electrical discharge through a gas or vapour
- Fluorescence or phosphorescence

Again the purpose of the agreement was defined so as to secure the co-operation of all parties in:

- Providing the most advantageous utilisation of their manufacturing facilities in producing lamps
- Securing and maintaining a uniformly high quality of product
- Arriving at a more economical arrangements for the distribution of the product
- Promoting more efficient methods of electric illumination and the increased use of light to the advantage of the consumer.

There was an understanding that all parties continue their business on independent lines to secure proper share of any increased business.

Detailed agreements were defined on the use of inventions, licences and the payment of royalties. Provisions were made for disagreements and the appointment of an arbitrator experienced in matters relaying to patents. All decisions made by an arbitrator were to be be binding on the participating parties.

As with the Phoebus Agreement, it was important to establish the relative position of the participating parties. For this purpose, territories were defined as:

1. Great Britain
2. Australia
3. New Zealand
4. India, Pakistan and Burma
5. South Africa
6. Remaining territories

The local meetings defined the local percentages, for the several fiscal territories. This in itself led to a complex and, of necessity, a carefully defined structure for the overall organisation. Again the framework was taken from the Phoebus organisation. There was an executive committee with a Chairman and Vice-Chairman, auditors and several advisory committees were appointed, for example:

- An Accountants Committee
- A Schedule Committee (translating lamps into units - statistics)
- A Juridical Committee (advising on legal matters)
- A Standardisation Committee
- A Technical Development Committee

Naturally the key aspect related to *'Excess and Deficit Sales'*; this related to the possibility of a member being temporarily unable to sell their agreed market percentage or indeed to oversell their quota. This was referred to as the Local Participating Percentage (LPP). Hence it was agreed that any party to the agreement selling above the LPP would share the profits resulting from these sales with the companies that had a deficit. A strict formula was established for this calculation relating to the agreed-shared profit per unit. This was based on the average realised price less 10% less the weighted average manufacturing cost per unit.

The average realised price was based upon the actual invoices less all discounts, rebates, credit notes, duties and taxes. For the average manufacturing cost, a remarkably detailed structure was established and agreed covering virtually all aspects of a cost price

Name of Member	Brand	% Share of ELMA Market	Financial Inter-connections
British Thompson Huston (BTH) Ediswan Swan Electric Metropolitan-Vickers Electrical	Mazda Royal "Ediswan" Cosmos Metrovick	( ) (33) ( ) ( )	Subsidiaries of AEI
General Electric Company (GEC)*	Osram	30	Cryselco
Philips Electrical Stella Lamp Co	Philips Stella	(10) (10)	Subsidiaries of Philips Holland
Cryselco	Cryselco	4	Jointly owned by GEC & Philips Holland
Siemens Electric Lamp & Supplies	Siemens Sieray	(10) ( )	
Crompton Parkinson		12	
British Electric Lamps (BELL) Aurora Lamps		( ) (1) ( )	

** GEC subsidiary, Pope's Electric Lamp Co. Ltd. with the brand 'Elasta', was not a member of ELMA.*

Figure 38: Members of ELMA and their market share in Great Britain 1950[11]

build-up including not only raw materials and labour, but all details of overhead expenses plus depreciation of plant and equipment.

The essence of this agreement was co-operation and trust, and little was left for ambiguous interpretation. Indeed not only were most eventualities anticipated, but also many of the words and phrases used in the agreement were defined, i.e. L.P.P. Quantity, Allocation Quantity, The Over and Under Delivered Quantity, The relief Quantity and Allocation.

At the end of each fiscal period all effective sales by all parties were converted into units by an agreed formula and if necessary modified each new fiscal period, in accordance with the latest conveniently available data on selling prices and types of lamps sold.

For Great Britain only, the businesses of B.T.H., Cryselco, Ediswan, G.E.C., Metrovick and Siemens were considered as one party. Similarly, outside Great Britain, the businesses of B.T.H., Ediswan and Metrovick were also considered as one party.

The General Meeting, consisting of the participating companies, was considered as the main deciding body but no decision was considered binding unless it had 90% of the votes cast.[12] There was an agreed procedure for setting up Local Meetings in each of the following Fiscal Territories:

Great Britain
Australia
New Zealand
South Africa

and for each of the following countries or groups of countries:

India
Pakistan
Burma
Ceylon
Malaya
British West Indies
Other Remaining Territories

These Local Meetings overlooked all the special interests in their territory and in many respects acted autonomously on local matters.

An Executive Committee with and Vice-Chairman with some

administration offices along with independent auditors were established, as well as Advisory Committees, where necessary.

Given the rather global nature of the Philips Lamp business, it was specified that Philips should appoint separate and independent accounts to confirm that lamps manufactured by or for Philips were sold in the territories concerned.

Specific conditions were laid down to cover the death of a business owner, sale/transfer, or the bankruptcy of one of the members' businesses. In each case, the new owner was required to undertake all the obligations of that member, or the General Meeting would have the right to veto such transfer. In other words, the General Meeting had very far-reaching powers.

Sales of lamps between companies belonging to the same group were not covered under this Agreement:

Philips and Stella,
BTH and / or Ediswan with Metrovick,
BTH, Cryselco, GEC, Metrovick and Siemens in respect of Great Britain only.

The period of this agreement, formally called 'The 1948 Lamp Agreement' was specified to commence on the 1st July 1948 and to remain in force until 30th June 1955, but it could be terminated any year on the 30th June with three months notice to all parties.

There were four annexes covering in detail:

1. Fiscal territories. (These were trading territories, not necessarily political boundaries.)
    - Great Britain including the United Kingdom, Northern Ireland, Eire, Channel Islands and the Isle of Man.
    - Australia. With all adjacent islands including the

Solomon Islands (Southern Group)

- New Zealand including: all immediate surrounding islands, Cook Islands, Kermadec Islands, Manihiki Islands
- India, Pakistan and Burma including the neighbouring islands
- South Africa including: Prince Edward and Marion Islands, North and South Rhodesia, Basutoland, Bechuanaland Protectorate and Swaziland
- Remaining Territories. This detailed all other parts of the then Commonwealth.

In the formal Agreement, all the above territories were actually specified in detail covering the name of every geographical location within that territory.

2. Provisions for calculating the weighted average manufacturing cost per unit for Great Britain.

In this annex of the Agreement, very detailed specification ensured as accurate a representation of the real situation as possible.

3. Local Participating Percentages[13]

Parties	Great Britain	Australia	New Zealand	India Pakistan Burma	South Africa	Remaining Territories
Philips	9.09246	48.57497	33.74575	35.62008	36.89245	20.8013
BTH/Ediswan		18.16496	19.1096	11.55292	16.1764	13.79285
British Block	77.56517					
GEC		12.13994	19.47455	27.94631	19.56835	36.97112
Metrovic		2.81754	2.74643	8.22438	5.10276	5.49247
Siemens		1.92529	1.8767	1.80421	1.85329	1.99483
Stella	0.92426					
Crompton Parkinson	12.41811	16.3773	23.04697	14.8521	20.40675	20.9086
Total	**100**	**100**	**100**	**100**	**100**	**100**

Voting Rights in the General Assembly

	No. of Votes
BTH	28,700,110
Crompton Parkinson	30,049,991
Cryselco	5,412,746
Ediswan	18,297,843
GEC	59,369,635
Metrovick	7,717,391
Philips	37,784,988
Siemens	20,007,110
Stella	1,496,469
BTH / Ediswan	10,225,668

4. All agreements governing the ELMA patent pool were fully detailed, particularly concerning GEC, BTH and Siemens with Ediswan, Metrovick and Cryselco.

Crompton, Aurora and BELL had been licensed by GEC, BTH and Siemens but their exports were restricted and Aurora and BELL had no patent protection outside the UK.

It was also interesting to note that both GEC and BTH had agreed to restrict sales of patented lamp parts and lamp making machinery to licensees.

GEC, BTH and Siemens had established licences in 1937 with British Luma (an associated company to the Swedish Luma), who had established a manufacturing unit in Scotland. This covered not only technical agreements but also quantities of lamps for sale in the UK and Ireland with specific arrangements covering the emergency supply to the Swedish Luma.

ELMA had extensive agreements with distributors, retailers and large users. These agreements regulated the sale of all ELMA lamps in the United Kingdom under the members' own brands, with the exception of flashlamps and cold cathode (neon) lamps.

The rules for distributors detailed that all purchase and resale transactions must be at the established prices and within the rules, terms and discounts laid down by ELMA. They laid down that there should be no payments or allowances, either direct or indirect, that could reduce the selling prices fixed by ELMA. Indeed a list of practices that was regarded as constituting inducements by either members or distributors that could lead to price cutting included regulating the scope of advertising, the form of quotations, invoices, contracts and tenders to be used, the charging of purchase tax, the supply of free samples, supplies on a sale or return basis, delivery and packing charges. Hence by this strict control anything that could be remotely regarded as price variation was prohibited, and the distributors were regulated by ELMA in the smallest of matters that might be constituted as competition with another distributor.

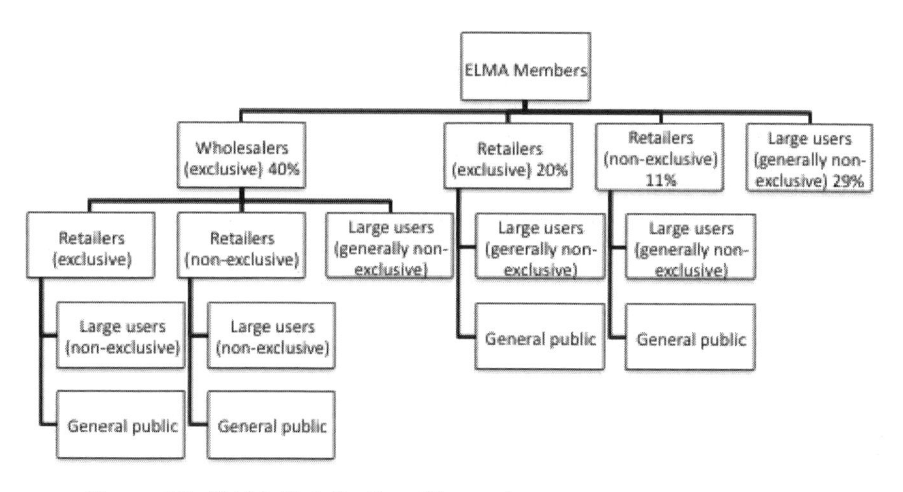

Figure 39: ELMA Distribution Channels

The Council of ELMA fixed the prices, selling conditions, discounts, and so on that were common to all members, for each type of lamp sold. A new or special lamp could be made only with the approval of ELMA and the price had to be fixed before the lamp was offered for sale. These prices represented the 'list' or retail selling prices to the general public. Discounts including rebates were allowed but on a strict fixed scale, according to the value purchases of ELMA lamps during the year. This meant that the net price charged to any given purchaser for a given type of lamp is laid down and is the same no matter which member made the lamp or who was the intermediate seller.

The discount terms authorised by ELMA varied in the first place according to the type of lamp. There were nine groups of lamps:

1. Incandescent lamps in 25 to 260 volts for general lighting service (GLS) purposes (except those in Group 8), projector and sign lamps in all voltages and traction lamps.

2. Lamps under 24 volts included all automotive, bus and train lamps, railway signal and other battery lamps (except flashlamps,).
3. Cycle dynamo lamps.
4. Lamps for miners' safety helmets.
5. Christmas tree decoration lamps.
6. Telephone switchboard lamps.
7. Lamps for general lighting service (GLS) purposes, known and marked 'Type B' now more commonly known as second brand.
8. Electric discharge lamps.
9. Radio panel lamps.

	Groups 1,8 & 9 *(incl. all general service) & discharge lamps)*	Group 2 *(Motor-car lamps)*
	%	%
Wholesaler	33 ⅓	47 ½
Retailer with exclusive agreement	25	38
Retailer without exclusive agreement	20	30

The users were split into 4 categories, each with their own terms; even outside this category there were users who negotiated their own terms.

Besides this discount, a rebate was allowed to some buyers based on their annual purchases. These rebates ranged between 1 and 12½ %.

These discount structures were detailed and comprehensive, and there was little deviation as it was very much a manufacturers' market. Their association, ELMA, controlled the commercial environment fairly strictly and there were too few large manufacturers outside the ELMA structure to influence this strict trading structure. It was most probably this along with relatively few large manufacturers that not only controlled the terms in the market but also controlled the supply of key components that prompted the government of the day to investigate the industry.

There were several manufacturers both large and small who were not members of ELMA at this time. The two largest were Thorn Electrical Industries (brand - Thorn) and Ekco-Ensign Electric Ltd. (brand - Ekco-Ensign). The British Luma Co-operative Electric Lamp Society was classified as independent, but it did have a special relationship with ELMA through its patent licence agreement with some of the members of ELMA. Several other small independent manufacturers existed, namely Maxim Lamp Works, Luxram Electric, Insular Electric, and Kingston Lamp Co.

The following table taken from manufacturers' returns made in 1950 to the Ministry of Supply and gives an approximate split of the share of ELMA and independent manufacturers.[14]

The total sales value at manufacturers' selling prices of the above output was £13.25 million and the estimated employment was 17,500 people. GEC and the AEI Group dominated the industry at that time with over 50% of the sales and employment.

Class of Lamps	ELMA Members (000's)	Controlled Companie (000's)	Independent Manufacturers (000's)	Total
Filament Lamps				
Flashlamps	26,586 (53%)		23,686 (47%)	50,272
Motor Car	25,983 (57%)	8,919 (19%)	11,718 (24%)	45,980
Other Filament Lamps				
Up to 24 volts	9,937 (69%)		4,514 (31%)	14,451
Over 24 volts*	85,651 (63%)	23,747 (17%)	27,035 (20%)	136,433L
Total Filament Lamps	148,157 (60%)	32,666 (13%)	66,313 (27%)	247,136
Discharge Lamps				
Fluorescent Lamps	2,843 (59%)		2,014 (41%)	4,857
Other Discharge	389 (100%)			389
Total Discharge Lamps	3,232 (62%)	2,014 (38%)		5,246

* This class includes all general lighting service (GLS) lamps

Figure 40: ELMA Market by Lamp Category 1951

Electric Lighting Fittings Association (ELFA)

In 1926 some of the lighting fittings manufacturers formed the Electric Lighting Fittings Association (ELFA) and the main objective of this association was *'by combination and co-operation to improve the status of the electric fittings industry, to eliminate undesirable practices and to promote the use of metal fittings.'* The early fittings industry was largely made up of many small manufacturers without clear standards and selling with questionable practices. This Association was to bring some structure to the Lighting Fittings industry.

The British Lighting Council (BLC)

In 1958 the BLC had been formed with the objective of promoting 'full and proper use of electric lighting in the service of the community'. The Council grew out of the Lighting Service Bureau, an organisation sponsored by lamp manufacturers that had functioned since 1924 and had established itself as a valuable source of advice and specialist knowledge on all lighting applications. Though strictly not a trade association, it was used as a vehicle to promote the industry on behalf of its members.

The BLC sponsorship was much more widely based, and it represented a complete cross-section of the lighting industry. It was backed by the British Electrical Development Association, the electric lamps industry through the Electric lamp Industry Council, by the lighting fittings manufacturers through ELFA and by the Electrical Contractors Association (ECA). Hence the BLC also provided a valuable liaison between supply, manufacturing, retail and installation interests in lighting.

During the 1960s the BLC rapidly gained recognition for its experts in various specialist lighting fields. Its activities included the organisation of exhibitions, conferences and lectures, and the issuing of publications on most aspects of lighting; it also provided a valuable information service and advice on lighting legislation.

Lighting Association (LA)

In parallel, another association had been formed in 1939 to represent small decorative lampshade manufacturers, Lamp Shades and Standards Association. They were formed to deal with trade from overseas competition, which had been fierce because of low labour costs in overseas markets. The Association had also purchased silk parachutes from the government so that lampshades could be produced. This association underwent many changes:[15]

1947	The Lampshades Manufacturers Association
1960	The Decorative Lighting Association
1968	The Decorative Lighting Manufacturers and Distributors Association
1971	The Decorative Lighting Association (DLA)
1991	The Lighting Association Ltd. (LA)

Its membership not only includes domestic lighting fittings manufacturers but also the major commercial fittings (luminaires) and lamp manufacturers. Indeed, importers are also included in their membership as the LA provides testing facilities for both lamps and fittings.

The LA has established a good lighting laboratory, which is widely used to test not only members' luminaires but a wide range of luminaires on the market, enabling the organisation to advise its members, trade and consumers on standards of a large variety of luminaires. The organisation provides a lobbying activity for its members, communicating information to government and trade alike.

Lighting Industry Federation (LIF)

Competition and some duplication between the various trade associations has always existed, even in the late 1960s. ELFA had been originally formed in 1926 to represent the lighting fittings industry and it continued to do so until its amalgamation in 1969 with ELIC to form the LIF.

Once this organisation was established, the continuity of the BLC also was questioned and it was decided to bring this under the same umbrella organisation.

During the 1960s the lighting (lamps and fittings) industries underwent great changes which ultimately had to be reflected in their trade associations which represented them. Many of the larger companies were members of both ELIC and ELFA (Electric Light Fittings Association); clearly there was duplication of functions and effort to represent the industry. This resulted in the two trade associations being amalgamated and a new association being formed – The Lighting Industry Federation. The main change is that the lamp standardisation has become totally internationalised, as the main lamp manufacturers are large international companies in which most products have global specifications. Variations of voltage are reflected in the operating control gear in the case of fluorescent and discharge lamps; as always, this is achieved with appropriately designed filament in incandescent and halogen lamps.

Hence, the role that LIF has played within the industry has continued to develop, particularly over the last 40 years. There has been less concentration has on the product standards, which have become more international, and more on environmental issues, these being established by International Electrotechnical Commission (IEC) and the European Standards Organisation (ESO). These organisations are largely staffed by the national standards organisations, for example BSI (British Standards Institution). Health and safety and application standards are still predominantly driven by international requirements. There continues to be a clear need for input at the formation and changes to legislation that requires the industry to put its views forward to the government and other official bodies. This role is carried out by the LIF.

As product standards became much more internationalised, the key roles of LIF also developed not only in providing a unified face to the government, the trade and consumers, but also in training and information on products/product standards. This takes the form of conferences, training seminars, etc., enabling the industry both lamps and lighting fittings (luminaires) to communicate with its total customer base.

Lighting Industry Association (LIA)

From 1st January 2012 the Lighting Industry Federation and the Lighting Association have merged with its mission the strengthen the industry and promote the benefits of good quality lighting by representing all aspects of UK, EU and international legislation and standards whilst protecting the interests of the public and members. This is a logical move and will avoid duplication of industry effort.

[1] The Electric Lamp Industry by Arthur A. Bright (Macmillan, New York 1949)
[2] HMSO Report on the Supply of Electric lamps (The Monopolies and Restrictive Practices Commission - 1951)
[3] Findings and Decisions of a Sub-Committee appointed by the Standing Committee on Trusts (1920)
[4] The Anatomy of a Merger by Robert Jones and Oliver Marriott (Jonathan Cape Ltd, London - 1970)
[5] HMSO Report on the Supply of Electric lamps (The Monopolies and Restrictive Practices Commission - 1951)
[6] HMSO Report on the Supply of Electric lamps (The Monopolies and Restrictive Practices Commission - 1951)
[7] HMSO Report on the Supply of Electric lamps (The Monopolies and Restrictive Practices Commission - 1951)
[8] HMSO Report on the Supply of Electric lamps (The Monopolies and Restrictive Practices Commission - 1951)
[9] HMSO Report on the Supply of Electric lamps (The Monopolies and Restrictive Practices Commission - 1951)
[10] HMSO Report on the Supply of Electric lamps (The Monopolies and Restrictive Practices Commission - 1951)
[11] HMSO Report on the Supply of Electric lamps (The Monopolies and Restrictive Practices Commission - 1951)
[12] HMSO Report on the Supply of Electric lamps (The Monopolies and Restrictive Practices Commission - 1951)
[13] HMSO Report on the Supply of Electric lamps (The Monopolies and Restrictive Practices Commission - 1951)
[14] HMSO Report on the Supply of Electric lamps (The Monopolies and Restrictive Practices Commission - 1951)
[15] The Lighting Association

6
The Controlled Companies

Almost from the outset, the Phoebus organisation experienced considerable competition from cheaper lamps made outside Europe, and it became its policy to meet this competition by acquiring some of these factories on the Continent to make and sell cheap lamps bearing special brands in the territories affected.

Special committees of the Phoebus organisation were set up to direct these fighting companies: the first was the 'Meteor' Committee; and after 1932, the appropriately named 'Hydra' Committee, on which two of the six members represented the British parties. Until the introduction of a tariff in 1932, the UK was a market for these fighting companies, one of which was N.V. Splendor of Nijmegen in Holland, whose subsidiary was Splendor, which was then a selling company only. Under the rules of the Hydra Committee, any local Meeting of the Phoebus parties could decide when Hydra lamps should be sold in their territory at a loss. There was always a conflict between the parties on the two aims, those who wanted to fight competitors and those who wished to see a return for the substantial investment made in these businesses.

As mentioned earlier, in the early 1930s GLS lamps were being imported into the UK, mainly from Japan and being retailed in a number of chain stores at prices often less than one half of the ELMA prices (6d. –1s.0d. versus 1s.7d – 1s.9d.)[1] The most notable were those being sold in the Woolworth chain at 6d. In 1935, the Phoebus organisation equipped Splendor to manufacture cheap lamps in the UK to compete with Japan; this was the first Hydra factory in the UK. This development was financed by N.V. Splendor, which was preferred by the British Group, who thought it would give it a more neutral character. The technical management was placed in the hands of the British Group, who guaranteed a minimum sale in the UK. Commercial control was in the hands of the Hydra Committee in consultation with the British Group, who could authorise sales below cost provided it refunded the loss. After the first year of operation of the new factory, it then traded at a profit.

It had been hoped that Splendor would be able to compete successfully with the Japanese in supplying lamps to Woolworths; however, Britannia, then independent, obtained the business and became the largest independent lamp manufacturer in the UK. In 1936, Britannia was acquired by Ismay Industries, who owned another lamp making company, John Ismay & Sons Ltd but, as mentioned earlier, by 1938, the main ELMA members acquired the whole of the lamp-making interests of Ismay Industries. The purchase was substantially in the form laid down by the Phoebus Agreement and was made only after the approval of the Phoebus organisation had been obtained. A management committee of three was formed, on which both GEC and BTH were always represented. When Splendor joined the group it was managed in the same way. The approval by the Phoebus organisation was given in order to avoid the businesses falling into the hands of 'non-members'. The British group explained to the Phoebus organisation that it was proposed:

> *to continue these businesses as independent concerns entirely outside the ELMA organisation and to use them to supply that*

> *section of the lamp market in Great Britain and the British Overseas Empire which is not prepared to accept the high quality Phoebus lamp at the Phoebus price.*

It was also said that:

> *in addition to retaining the business they now hold, these two companies in co-operation with the British Splendor Company will succeed in taking business away from the remaining outsiders and thus controlling this section of the market with the least possible injurious effect on the regular Phoebus business.*[2]

The companies were expected to yield a fair return on capital invested and sales were not to be reckoned against the quotas of ELMA members. Their aim to control the market for cheaper lamps with the least injury to their own high priced brands and to use these companies, if required, as fighting companies, was to last for over thirty years.

The main ELMA members (exceptions being Stella, BELL and Aurora) had purchased the majority of the share capital of Britannia, Ismay, Splendor and Evenlite. Gnome was a wholly owned subsidiary of Britannia. Splendor had been a wholly owned subsidiary of N.V. Splendor of Nijmegen. This at the time was 40% owned by N.V. Philips, who later acquired the whole of the shareholding in N.V. Splendor and then sold the share capital of Splendor UK to the ELMA members holding the rest of the share capital of the Controlled Companies.

At about the same time, the ELMA members introduced limited quantities of lamps of lower quality to their main brands, which were to be known as 'Type B' and to be retailed at 1s.0d. The design life was set at 900 hours (normal standard 1000 hours) and there was a system of fines for longer life and luminous efficiency above the limit laid down.

In the late 1930s, this move helped ELMA members to regain control of the cheaper lamp market, but after the war these companies were operating more independently. The Controlled Companies referred to in the ELMA structure were not themselves members of ELMA but were controlled by ELMA members. These were:

Name of Manufacturer	*Brand*
Britannia Electric Lamp Works Ltd	Britannia
Ismay Lamps Ltd	Ismay
Splendor Lamp Co Ltd	Splendor
Evenlite Tube lamp Development Ltd	Evenlite
Gnome Electric lamp Works Ltd	Gnome

Their relationship with the ELMA members can be seen in Figure 37 The ELMA Structure 1949 on page 163. Practically the whole of the share capital of Britannia, Ismay, Splendor and Evenlite was jointly owned by the ELMA members (except Stella, Bell and Aurora), Gnome was a wholly owned subsidiary of Britannia.

Evenlite was always in a special position, as it manufactured architectural and other filament lamps; it applied to join ELMA in 1946, but after negotiations, ELMA members acquired Evenlite and updated their equipment and concentrated their production of these types of lamps in the Evenlite factory under their own brand names. BELL was the exception as they continued, independently, to manufacture some of these lamps.

Ismay and Britannia concentrated on mains voltage general

lighting service (GLS) lamps, whilst Splendor made both GLS and motor-car lamps. Gnome specialised in motor-car lamps and did not get involved in GLS lamps. None of the Controlled Companies was involved in the manufacture of discharge lamps.

These companies had approximately 35% of the market before 1939, but following the war they lost some share and by 1950 had less than 30%. Their main customer continued to be the F. W. Woolworth chain store. Almost the whole of the Britannia and much of the Ismay production went to service this account.

Woolworths was the main retail outlet for GLS lamps up to 1970; in both the pre- and post-war period they were the benchmark for low prices:

		1939 price	*1950 price*
ELMA (main brands)	60w clear	1s 9d (8.75p)	1s 1d (5.4p)
	60w pearl	1s 7d (7.9p)	1s 1d (5.4p)
Woolworths	60w clear	6d (2.5p)	8½d (3.5p)
	60w pearl	6d (2.5p)	9¼d (3.9p)

The Controlled Companies fixed retail prices and trade terms for lamps sold under their own brands but had no arrangements to enforce these prices. Lamps marked with customers' own brands or unbranded were sold at negotiated prices and the terms of resale thereafter were the responsibility of the retailer. Whilst retail outlets were their main customers, they also sold lamps to Government departments.

Indeed, in the 1951 Monopolies Commission Report, it was stated that the Controlled Companies had been left to run their businesses independently of the shareholders' management committees and that there was no evidence that they had been used as fighting companies. On this issue the 1951 Commission said:

> *The parties to the Phoebus Agreement maintained fighting companies as another means of attacking independent manufacturers. One of the objects of the ELMA members in buying Britannia and Ismay was to get control of the market for cheap lamps and to use these companies with Splendor as a means of 'taking business away from the remaining outsiders', that is to say as fighting companies. The evidence is, however, that these companies have in fact been left to run their business largely independently of the shareholders' management committees. There is no evidence that they have been used as fighting companies and we are assured that there is no present intention of so using them.*

There was one member of the Commission who did not accept this; and in an addendum to the report it considered that it was unrealistic to expect the Controlled Companies to compete freely and effectively with ELMA members while still being owned by them, and recommended that the Government should acquire the share capital of these companies, but this was never pursued.

In 1954, Thorn started to supply GLS lamps to Woolworth through an agency of the Universal Distributing Co. Ltd. under the brand name 'Vesta'. Britannia sold only to a very few multiple retail customers and aimed not to sell to buyers who would be in direct competition with Woolworth. Despite this loss of business, Ismay and Splendor significantly increased their sales, thus helping to keep the Group in profit. There were some small outside shareholders in Britannia and, these being private companies, they presented problems, as their expectations were different; and steps were taken in 1965 to acquire these outside shares. Complete control was finally acquired in 1966.

There was little or no major capital expenditure up to 1963; indeed, most of the equipment dated back to the 1930s, and the period from the mid-1950s had seen significant improvement in lamp- making equipment. The shareholders had invested in faster, much more efficient machinery in their own factories but they were not inclined to make such capital investments in the Controlled Companies. In 1963, a proposal was made to completely reorganise with a view to cutting costs, and it was proposed that the partners could make a proportion of the production. This included the concentration of GLS production at Britannia's works at Park Royal, the concentration of other types at Splendor's works at Wimbledon and the closure and sale of the Ismay's factory at Ilford. A maximum practical production of 36 million GLS lamps was assumed compared with the then capacity of 27 million, with a substantial increase in sales forecast. However, with one exception, the Shareholders Management Committee was not in favour of the Controlled Companies' production continuing at the current level. They considered that a proportion of the Companies' requirements of lamps could be supplied by the shareholders and that the purchase of any new equipment should be very limited. Indeed, AEI itself was undergoing some reorganisation due to falling profits, and wanted to seize this opportunity to increase its own production, as it had lost a significant sales volume in the previous five years. Whilst Osram was in favour of the new proposals, GEC were reluctant to spend money on new equipment. Philips and Crompton were running their own cheap lamp businesses independently, partly because there was still some production capacity available, and new equipment could give the Controlled Companies more up-to-date facilities than in their own factories. Philips and Crompton were urged to integrate their cheap lamp businesses but both were reluctant to do so since their principal competitor, Thorn, had its own cheap lamp business. Although opinion was divided on an approach to Thorn to integrate their cheap lamp business into the Controlled Companies, no actual approach was made.

Osram and AEI Lamps & Lighting indicated they were ready to move into the cheap lamp market independently and if necessary; all shareholders agreed this could mean the ultimate closing down of the Controlled Companies. In February 1964, Philips and Crompton indicated that they were agreeable in principle to giving up their respective second brands, Corona and Hygrade respectively if certain other brands were also brought within the scope of the Controlled Companies. Philips would not be willing to invest further unless such an arrangement was made; if it were not, it might withdraw altogether. Although Osram considered the Controlled Companies to be reasonably successful, it reminded the other shareholders that it and AEI L & L could combine and take all the production with them. AEI L & L stated it did not want to start a second brand, but would feel forced to do so if Philips and Crompton continued with theirs, and if Osram started one. In July 1964, with BLI now a shareholder through its subsidiary AEI L & L, along with the other shareholders, did not support any increase in GLS production. It was agreed that they had enough capacity to obviate the need for putting new plans into the Controlled Companies.

The GLS production capacity limit was set at 20 million, and additional requirements would be met by the shareholders on the basis of their respective shareholdings. However, the existing machinery had become too inefficient to meet the needed costs, and after much discussion it was agreed to purchase two new lamp-making groups. The Splendor factory at Wimbledon was closed and part of the plant transferred to the Britannia Works at Park Royal, where the new equipment would be installed, making Britannia the main manufacturing and selling company.

During 1965-66 profits and hence dividends improved, but in 1966, due to excess stock, borrowings increased, and by 1967 the Park Royal factory was becoming uneconomic for many reasons, including the availability of suitable labour. Park Royal was closed and the plant moved to the Ilford factory. Britannia became a wholly owned subsidiary of Ismay. This caused some changes in the shareholdings of the other companies. These shareholdings (held through nominees) were now:[3]

Ismay – (manufactured and sold GLS filament lamps)

	%
AEI L & L	43.7520
GEC	30.2795
Crompton	12.6310
Philips	9.2505
Stella	0.9393
Cryselco	3.1477

Splendor – (sold GLS filament and motor-car lamps)

	%
AEI L & L	43.9733
GEC	29.9183
Crompton	12.4184
Philips	10.0167

Cryselco	3.6833

Evenlite – (manufactured and sold – mainly to shareholders – architectural and tubular filament lamps)

	%
AEI L & L	43.975
GEC	33.600
Crompton	12.425
Philips	10.000

Ismay had the following subsidiaries

Britannia: had by now become a selling company for GLS lamps for the group. Britannia had two subsidiaries, Gnome Electric Lamp Works and Norma Electric Co., both of which were selling speciality filament lamps and motor-car lamps.

Union Lamp and Lighting Co. Ltd: wholesalers for GLS lamps to large users under the brand name 'Union'

Briton Lamps Ltd: originally a buying company but used for exported lamps.

CEL Ltd: for exported lamps only.

Machine Design and Construction Ltd: for the manufacture and sales of lamp-making machinery and parts.

By 1967 all the shareholders were in direct competition with the Controlled Companies. The only exception was Cryselco, which itself was jointly owned by GEC and Philips. The distribution market was changing, with the growth in the domestic sector coming from the self-service stores; this trend had already been seen in the USA. With all the main brand companies concentrating on a wholesale distribution policy, they were not satisfied to leave this sector in the hands of the Controlled Companies. BLI had its Omega organisation; hence, the other shareholders sought an 'arm's length' solution to meet this growing market sector without creating a confrontation with their main customer group, the wholesalers.

BLI: Omega had been acquired by Thorn in 1957. Astralec was formed in Thorn to handle the sales to Woolworth.

Osram: Ascot was acquired in 1965.

Philips: Corona was formed in 1961 but by 1967 had virtually ceased to trade. Controlling interest had been purchased in Luxram and Kingston in 1965 and 1966 respectively.

Crompton: Hygrade brand.

A further complication came from the technical construction of the filament lamp. In the mid-1930s the coiled-coil filament had been invented, giving the possibility of increasing the lumen efficiency (light output) by up to 10%. Previously the filament had been only a single coil. It was not until after the 1939-45 War, when the industry was re-investing, that machinery was designed to make coiled-coil filaments on a large scale and the lamp-making equipment adapted to assemble this filament in the lamp-making process. It was soon realised that this double coil being shorter created the opportunity to mount the filament mechanically, thereby eliminating one of the restrictions to increasing the speed of the lamp-making groups, which was previously limited by an operator feeding the filaments by hand and the inability of the equipment to handle the much longer single coil.

As the shareholders in the Controlled Companies had not wanted to re-invest, they had created the image that the main brand lamps were of higher quality and maintained a higher price for the main brand lamps, the implication being that the second brand being sold by the Controlled Companies and other companies not manufacturing coiled-coil lamps were inferior, therefore justifying the lower price. The self-service retailers in the consumer market, which were growing in size and wanted to purchase direct from a manufacturer, were offered these single coil lamps. The wholesalers were satisfied as they had the main brand lamps and their customers, the small electrical retailers, were not subject to the retail price-cutting being carried out by the multiple self-service retailers.

In this way, the decision to close the Park Royal factory and concentrate all the production at Ilford, along with the shareholders supplying up to half the requirements, led to further complications. The partners had to supply single coil lamps, which by this time were not only more expensive to make; their capacity on older equipment was more limited. Many larger retail customers were not able to purchase the main brand lamps at the prices equivalent to the second brand lamps, and have the freedom to set their own retail prices. Consequently, many of the larger multiple outlets purchased from the Controlled Companies and other second-brand companies with their own branded lamp.

In their presentation to the 1968 Commission, the shareholders argued that their reorganisation and limited investment in more modern equipment allowed the Controlled Companies to become more efficient. The main points of their presentation were

1. They, the Controlled Companies, would be able to regulate their purchases to suit their requirements more easily and

more economically than if they had to adjust their own production level to suit fluctuating demand.

2. The shareholders would be more tolerant towards cancellation of orders than would be the case if there were no shareholding interests. (By way of example, they showed that the Controlled Companies had cut back their requirements to 5 million in the year ending March 1968.)
3. They had an assurance of the supply of their requirements from the four shareholders together which was greater than it would have been if the Controlled Companies had been dependent on one or two external suppliers.
4. The Controlled Companies were at the same time placed under no economic disadvantage by not manufacturing their total requirement themselves. The transfer prices were based on the Controlled Companies' own factory cost prices.

Each of the shareholders had decided they wanted to compete in the so-called 'cheap' lamp market, which was really another name for the growing competitive consumer self-service market, in which the retail customers did not want to be restrained by manufacturers' recommended retail prices. In 1968, the shareholders were very active with the following brands in this market sector:[4]

BLI	Omega, Nura and Vesta (exclusively for Woolworth)
Philips	Corona, Luxram, Star, Kingston, United and Insular
Osram	Ascot
Crompton	Hygrade

In addition, they were supplying many own (customer) brands, so the function of the Controlled Companies to compete in the 'cheap' lamp market was no longer just against the independents but with the individual shareholders.

When the Monopolies Commission was set to abolish the recommended retail prices in 1968, which would lead to the major customers of the Controlled Companies being able to set their own prices, it left the future of this organisation in question. The Commission said in its 1951 Report that:

> *The preservation of the Controlled Companies as competitors would become even more important if ELMA or its members were to absorb any of the independent manufacturers. It recommended that, if that were to happen, or if the position of these Companies ceased to practise an independent price policy, the position of these Companies should be reviewed in the light of the situation as it would then exist.*

The possibility envisaged by the 1951 Commission that the major manufacturers might acquire several independent manufacturers had now occurred and in the field of GLS lamps, there were only two independent manufacturers left British Luma and Maxim. The former concentrated almost entirely with sales to co-operative societies and the latter being very small. There were a growing number of independent manufacturers of fluorescent lamps but, although this market was growing, the Controlled Companies did not make these lamps and their ad hoc requirements were purchased from independent manufacturers.

In the summing up the 1968 Monopolies Commission were very perceptive in the comments made:

We consider the Controlled Companies, which are responsible for over half the cheap lamp market, although their share of the total market is only about 3%. They are now something of an anachronism. They were set up or acquired before the war for the purpose of combating cheap lamp competition, especially from Japan, and of securing control of the cheap lamp market with the least injury to the higher priced brands. Already in our predecessors' report (1951), they had ceased to be used in their intended role as fighting companies.

Since then there has been a radical change in circumstances. With the growth of the cheap lamp market each of the shareholders has thought it desirable to participate in it directly, with the result that with the exception of one small company the cheap lamp manufacturers with which the Controlled Companies compete are now controlled by the shareholders themselves. From the point of view of the shareholders themselves, therefore the Controlled Companies' role has largely disappeared.

So the scene was set; the shareholders being confronted in public with their own internal thoughts were now ready to consider the closing down of the Controlled Companies. In 1969 the business, including brands, sales, plant and equipment, were divided up between the shareholders, and these companies ceased to exist.

Although it was not realised at this time, this loss of several hundred jobs was the start of a very difficult period for the lamp industry in the UK. The 1970s were to bring about further dramatic and very long-term changes within the lamp industry worldwide. The days of cheap energy were about to come to an end; and this was preceded in the UK with a self-inflicted energy crisis with a coal mining strike leading to huge electricity cuts, a three-day week, and the first real decline in demand for lamps, particularly incandescent lamps, since their invention. This was to lead to further industry rationalisation. The days of cheap oil were also about to end.

[1] HMSO Report on the Supply of Electric lamps (The Monopolies and Restrictive Practices Commission - 1951)
[2] HMSO Report on the Supply of Electric lamps (The Monopolies and Restrictive Practices Commission - 1951)
[3] HMSO Second Report on the Supply of Electric Lamps - 1968
[4] HMSO Second Report on the Supply of Electric Lamps - 1968

7
Lamp Components

There was always a high degree of vertical integration within the larger manufacturers, and this has been, and still is today, a feature of the lamp industry, forming a significant barrier to entry as it is where a major part of the technology exists. The larger firms continued to invest in improvement in the manufacture of glass (bulbs and tubing), particularly tungsten, but also all the metals needed for the lamp making process. This process was originally a hand assembly; later it became part hand and hand-operated machines; and today it consists of sophisticated automatic assembly machines.

Indeed, because of the major companies having international links, practices adopted in both America and continental Europe spread to the UK. Glass manufacture, while not in itself being a very sophisticated technological process, was very capital intensive. The real dedication and development was in the equipment to produce the glass components, i.e. the bulb or tubing. The larger ELMA members soon realised that the competitive edge was not in the component technology but in the cost. Glass required major capital investment, and the glass-processing machinery required continuous development to innovate, automate, and run at high speed to achieve low cost. Combining resources to produce low-cost bulbs and selling surplus capacity to other competitors could thereby produce a competitive edge.

During the early developments of the industry, there was a wide diversity of practice among the lamp manufacturers as to the components made for them or bought elsewhere. In the period up to 1918, the British manufacturers were highly dependent for their supplies of components on imports or, in the case of glass bulbs, on mouth-blown production. Following the 1914-18 War when a period of consolidation took place, particularly on the Continent, discussions were taking place about co-operation on the components involving high capital expenditure, i.e. glass bulbs and caps. Following the formation of Osram GmbH in 1920, Vereinigte Lausitzer Glaswerke AG in Weiswasser (Germany) was one of the factories taken over; discussions took place with Philips on the possibility of setting up a joint operation, not only for glass but also for caps. Although the principles were agreed, Anton Philips considered the situation in Germany where inflation was raging to be too uncertain for entering such far-reaching agreements with Osram. Hence, Anton Philips informed William Meinhardt of Osram that he preferred to wait until exchange rates settled down.[1] Whilst this plan lapsed, the basis of future co-operation was established.

The trade associations such as ELMA assisted in bringing some commonality in these activities. Between 1918 and 1939 the manufacture of components was developed on a considerable scale. The production of most lamp components on a commercial scale requires major investments in the manufacturing process. In glass factories, apart from the furnace, there are the bulb-blowing machines and plant for drawing the tubing and rod as well as the equipment for cooling (annealing), testing, and cutting into lengths. The British patent rights for the Westlake bulb-blowing machine, which originated in the USA, were acquired by a company sponsored by some ELMA members; and GEC and BTH started to produce machine-blown bulbs in their own glass works in 1919. They had the patent rights on this Westlake machine up to 1932. In 1925, these same two ELMA members, GEC and BTH, , formed a company to manufacture lamp caps called Lamp Caps Ltd. Also, during this period between the wars, the ELMA members significantly developed their wire drawing capacity. The non-ELMA members did not appear to significantly invest or develop the manufacture of components.

By 1939, the ELMA members were jointly self-sufficient, and sold very little to non-members, who relied mainly on imports from Europe. In 1939, when it became necessary to reduce imports, ELMA told the Board of Trade that as regards lamp components its members were 'probably in a position to meet the whole requirements of the country from British sources'. Hence, in 1940 the position changed radically when imports from Europe ceased entirely; during the war years the independent manufacturers became completely dependent on the ELMA members for their supplies of lamp components. From 1942 to 1945, the Government controlled the supply of lamp components, in particular glass bulbs, tungsten and molybdenum wire; and there was an allocation system among all manufacturers. Post 1945, restrictions on imports were relaxed but some prices were higher and specifications varied from those being produced by the ELMA members.

Patents formerly covered many components of filament lamps. An agreement in 1921, made between the two principal component manufacturing members of ELMA, GEC and BTH, required them not to sell patented components, other than glass, except to licensees. There was an agreed price for the sale of tungsten wire to licensees. The Phoebus Agreement, in 1925, contained a provision prohibiting the parties from giving 'aid' directly or indirectly to manufacturers who did not share 'the burdens and obligations' of the Agreement. The application of this provision to the supply of components was the subject of much discussion. The Phoebus organisation regarded the control of lamp components generally as one means of joint attack on the so-called 'outsider'. The parties sought to gain control of the independent sources of supply of components by acquiring them or making agreements with the owners and, secondly, to prohibit or limit the supply of components to non-parties to the Agreement. Clearly, this provision was not rigidly pursued and only applied from time to time when commercial pressures were sought. In the UK the Monopolies and Restrictive Practices Commission of 1951 found no evidence of its application. Nevertheless, agreements made by the Phoebus organisation with some continental suppliers of components did have a small effect on the British independent manufacturers who relied on imported components.

BTH and GEC also concluded an agreement with Philips (Holland) to exchange manufacturing information and research and non-exclusive cross-licensing in the UK; no royalty was to be payable on either side. Crompton maintained an arrangement it had made with Philips before the war but paid royalty in return for non-exclusive licences under the Philips patents.

Until 1948, British Luma was the only non-ELMA lamp manufacturer in the UK to have a patent licence affecting lamps from ELMA members. This went back to an agreement made with the Swedish Kooperativa Förbundet in 1937, which controlled a factory in Scotland making lamps and had an interest in British Luma; it paid 3% royalties for this privilege. Under the same agreement, the British Luma agreed to maintain prices and terms to all outlets except sales to co-operative societies, and to maintain their sales quota. It was reported to the Monopolies and Restrictive Practices Commission in 1950 that GEC and BTH had dissuaded the CWS (Co-operative Wholesale Society) from laying down a glass plant to make tubing for fluorescent lamps as they maintained there was sufficient capacity in the UK to meet all demands. This obviously led to favourable prices being negotiated from ELMA members. From 1948, the agreement with British Luma was maintained on an ongoing basis with an exchange of letters, but it was finally terminated in 1951.

Two leading manufacturers of fluorescent lamps, Thorn and Ekco-Ensign, had been negotiating with GEC and BTH for some years to obtain certain patents. GEC had issued writs against Thorn in 1945 for infringement of a patent relating to fluorescent powders.

The glass bulb-blowing machines in the 1950s in the UK mainly consisted of two Corning 'Ribbon' machines with an output to meet most of the UK requirement for standard glass bulbs. The company established for the manufacture of the glass bulbs was 'Glass Bulbs Ltd.' jointly owned by BTH and GEC. This had been one of the sources of constant complaints by the independent manufacturers that Glass Bulbs Ltd. discriminated both in price and times of shortage against them. Indeed, in response to the request by the Monopolies and Restrictive Practices Commission the following letter was sent to the Ministry of Supply in June 1951 by the acting Chairman of Glass Bulbs Ltd. in Harworth:[2]

Dear Sir,

We have been asked by the Monopolies and Restrictive Practices Commission to write to you amplifying the assurance given to me in a letter dated 20th June 1947 in connection with the discrimination of glass bulbs made on the Ribbon Machine at Harworth.

We again confirm that bulbs will be distributed, without discrimination, and that the prices will be charged will be the same – quantity for quantity – to all purchases in the United Kingdom, with the exception that the proprietors, viz.:- The British Thomson-Houston Company Limited and The General Electric Company Limited, will purchase at lower prices; the same applies to the controlled firms where we have an obligation to the shareholders.

I have sent a copy of this to the Secretary of the Commission.

In addition to the manufacture of glass was the manufacture of tungsten wire, molybdenum wire, copper wire, copper-clad nickel-iron wire and nickel wire, and finally lamp caps. The manufacturing processes for all these components required investment in research and development as well as major capital investment. Indeed, the technology in the production of both quality and low cost components was and still is today the major contributor to successful lamp making. A key to the larger companies' success was the management of a successful component activity. The manufacture of glass was a specialist technology requiring major capital investment, mostly bringing capacity greater than an individual manufacturer's requirement. This resulted in joint ventures being entered into; an example of this was Glass Bulb Ltd. in Harworth, Yorkshire, jointly owned by BTH and GEC. A similar enterprise had been set up by Philips (Holland) and Osram (Germany) at Emgo in Belgium to manufacture glass bulbs and tubing.

The policy of the Phoebus organisation was reflected in the patent licences granted by GEC, BTH and Siemens before the war. Aurora, BELL and British Luma as licensees undertook not to supply any lamp components, whether patented or not, to manufacturers not approved by the licensors. The patent licence to Crompton included filaments and tungsten wire and later fluorescent powders, whereas BELL undertook not to make filaments. This policy was not continued during the war, but the ELMA members established a new Lamp Agreement in 1948; its provision was very similar to the one contained in the Phoebus Agreement. Though it did not apply to glass, it did apply to filaments and fluorescent powders. It also continued the Phoebus policy of fixing common prices for the sale to:

a. Fellow-members,
b. The Controlled Companies
c. British Luma
d. Other independent manufacturers.

This policy was looked at very closely by the Monopolies and Restrictive Practices Commission in 1951. While it accepted elements, the Commission wanted clear allocation rules in times of shortage. On prices it recommended that members and non-members should be charged at equal prices quantity for quantity. It did accept that lower prices could be charged to the parent or subsidiary companies. In the case of bulbs made on the Ribbon Machine, the Commission required Glass Bulbs Ltd. to publish list prices and discounts at which it was prepared to sell.

The independent manufacturers complained to the Commission that the ELMA members had tried to hamper their manufacturing and sales activities by withholding supplies of essential components, particularly bulbs and caps, or by delivering the wrong types, for example the wrong size of glass bulbs. The documentary evidence submitted to the Commission did not fully convince them although they were satisfied that, in periods of shortage in the post-war period, unfair allocation of supplies had taken place.

There was some independent production of tungsten and molybdenum; a subsidiary of Thorn made lamp caps and no doubt there was some competition between the ELMA members. There was a virtual monopoly of glass bulbs by GEC and BTH who, working in close co-operation, had been producing 200 million machine-made glass bulbs at Glass Bulbs Ltd. in Doncaster; but in 1951 they installed the first of two Corning Ribbon Machines increasing the potential output to 450 million. These Ribbon Machines gave a saving of 15%–20%; they were to become the start of a new generation of bulb-making equipment that was to spread throughout the world. The Ribbon Machine's successors, faster and much more sophisticated, are widely used today. The Commission addressed this monopolistic capability in 1951, when drawing up its recommendations for pricing and allocations in time of shortage. However, the Commission did recognise that this new Ribbon Machine would create a virtual monopoly by Glass Bulbs under the direction of GEC and AEI Group; hence their recommendations that, in addition to the undertaking given in the above quoted letter to the Ministry of Power, Glass Bulbs should publish their prices and discounts.

The 1951 Commission also recommended that in times of temporary shortages of any materials or components, the ELMA members had an agreed formula for handling allocations between purchasers; in the absence of such a formula the Government would take direct responsibility by imposing control. ELMA's component-making members should keep in close touch with the responsible Government department in order to ensure that no misunderstandings arose concerning the fairness of the shares received by the independent manufacturers.

Component	**Price Index on Sale by ELMA Members to:**			
	Independent Manufacturers	British Luma	Controlled Companies	ELMA Members
Glass Bulbs				
Clear - without contract	100	85	73	100
- contract (5 million pa)	92.5			92.5
- contract (10 million pa)	90			90
Pearl - without contract	100	80	74.5	94
- contract (5 million pa)	92.5			87
- contract (10 million pa)	90			85
Caps				
- up to 2 million pa	100	93	93	93
- 2 to 7 million pa	96.5	89.5	89.5	89.5
- over 7 million pa	93	86	86	86
Tungsten and Molybdenham wire				
- without contract	100	90	87.5	90
- with annual contract	92.5	83	81	83

Figure 41: ELMA Members Component Selling Price Index in 1950[3]

With the additional volume created by the installation of the high-speed Ribbon Machines, the key to success was to ensure they were fully loaded; hence it became essential to sell the volume. Naturally, these put emphasis not only the home market, but also on export sales, so the bulb supply situation remained reasonably stable over the next couple of decades. When the 1968 Commission investigated the Industry, their findings concluded that the undertakings given by Glass Bulbs Ltd. to the 1951 Commission were largely maintained. Given that the standard bulbs for each customer were virtually the same, the pricing arrangements given to the Commission were:[4]

1. To parents BLI and Osram GEC, cost plus 6%
2. To Philips, cost plus 17.5% (based on 10 year contract)
3. To the Controlled Companies, list prices less 17.5% rebate
4. To all other customers, list prices less the following rebates:-

Quantity (Millions)	Rebate (%)
0 to 2.5	nil
2.5 " 5.0	2.5
5 " 10	5
10 " 15	7.5
15 " 20	10
20 " 30	12.5
30 and over	17.5

The business situation continued to be dynamic; the first complaints were made by the independent manufacturers against the two cap suppliers. Lamp Caps Ltd. being jointly owned by GEC, and AEI having been set up in 1922, they made the standard BC (B22) and ES (E27) plus a some special types. Lamp Presscaps, a subsidiary of Thorn until 1964, also made a range of the standard caps, but they had concentrated on the fluorescent bi-pin caps and become the sole supplier of these caps. These two companies produced virtually the whole of the UK requirement. Only a relatively few caps were imported; these were mainly special types from NV Vitrite, a subsidiary of Philips.

While discussions had taken place on several occasions on the amalgamation of Lamp Caps and Press Caps, it was not until the formation of British Lighting Industries (BLI), the amalgamation of the lighting interests of Thorn and AEI, that close collaboration became apparent. In 1966 it was finally agreed to concentrate all vitrited (the internal glass insulation) caps at Chesterfield. This was achieved through the purchase by Lamp Caps of the Lamp Presscaps' business in this type of cap; and in return, Lamp Presscaps acquired all the production of bi-pin fluorescent caps at Edmonton. This became effective in January 1967, effectively creating the dominant/monopoly situation that was closely investigated by the Monopolies Commission in 1968, given that an undertaking had been given to the 1951 Commission not to differentiate between customers and to operate two price lists, one for parents and the other for all other customers based upon standard prices and predetermined quantity discounts. In 1957 this system was abandoned and the parents took supplies at cost plus 5%, and other customers used the initial price previously agreed less the appropriate quantity discounts; but these varied by negotiation.

This pricing system gave rise to several complaints and adverse comments from customers and competitors alike. With imported caps having to carry the burden of import duties, the home producers were protected and a virtual monopoly situation existed, which the 1968 Commission said acted against the public interest. Consequently, the 1968 Commission made two key recommendations: first, that Lamp Caps and Lamp Presscaps should charge the same price to their parents as to other customers with the additional price paid being recovered in the additional profits and secondly, that quantity discounts should be only such as could be justified by actual variations in cost.[5] Strict undertakings had to be given to the then Board of Trade.

There is little doubt that the break-up of the Controlled Companies and the Second Report on the Supply of Electric Lamps (The Monopolies Commission) was the start of the break-up of the monopolistic trading practices of lamp components industry. During the 1970s, a gradual increase of lamps and components from the Far East made the continuing fixed price policies on components unrealistic, and the effect of the significantly lower manufacturing costs put the whole industry under cost price pressure.

In the 1980s and 1990s the industry commenced major changes. As mentioned earlier, GE of America took over Tungsram in Hungary and the Thorn lamp organisation, and gradually moved the Thorn lamp base, including all the component activity, to Hungary. Already GE was sourcing many of its components from low-wage countries in Eastern Europe and the Far East.

Osram GmbH finally managed to take over GEC Osram, which underwent a similar fate to the Thorn lamp activity, particularly as much of the GEC lamp manufacturing had suffered from under-investment and outdated machinery. Therefore, again, GEC lamp activity was gradually integrated in the Osram GmbH manufacturing plants. Osram GmbH, like GE, was also investing in manufacturing in low-wage countries; and even more recently, they took over Svet in Smolensk in Russia in 2004 .

Philips Lighting was also moving much of its older manufacturing to low-wage countries in Eastern Europe and the Far East. In 1991, Philips took over Polam-Saw in Pila, Poland and invested heavily in component manufacturing there. In addition, similar major investments were being made in the Far East, particularly in China and Indonesia.

The result of all this industry restructuring was the end of most joint component manufacturing in the UK. Within Europe, Osram and Philips continued with a joint venture in glass technology started in 1966, but the product portfolio volumes were significantly reduced, and the product ranges increased, for example glass for the solar heating and photovoltaic markets ?

Coincidently with the new millennium came the real decline in demand for incandescent lamps; and manufacturers who were unsatisfied with the component pricing arrangements went to Eastern Europe or the Far East for their component supplies. With the larger manufacturers setting up their own manufacturing facilities in low-wage countries the traditional inter-company component supply all but disappeared. Furthermore, many distributors, both retail and importers, were bringing in large volumes of incandescent low-cost manufactured lamps.

[1] Minutes of the meeting of the Supervisory Board, 7.9.1922: Philips Corporate Archives

[2] HMSO Report on the supply of Electric Lamps (The Monopolies & Restrictive Practices Commission) 1951

[3] HMSO Second Report on the Supply of Electric Lamps - 1968

[4] Second Report on the Supply of Electric Lamps 1968 (The Monopolies Commission)

[5] Report on the Supply of Electric Lamps 1968 (The Monopolies Commission)

8
The Dynamic Sixties

Following the 1939–45 war and the extensive enquiry into the lamp industry by the Monopolies and Restrictive Practices Commission in 1949-1951, a period of stability was maintained for a few years. The estimated market shares in 1951 were:[1]

	Per cent
AEI Group	33
GEC	30
Crompton	12
Philips and Stella	10
Siemens	10
Cryselco	4 (jointly owned by GEC and Philips)
BELL/Aurora/Others	1

As with all industrial or commercial industries, this stability was not to last. Probably one of the most significant changes was the continued growth of Thorn Lighting and the relative weakening of AEI in the lamp field. AEI had some of the most prominent businessmen of the period as executives. Viscount Chandos, later Lord Chandos, was re-elected as Chairman in 1954 following a period in politics. In 1955, the following year, AEI bought Siemens, giving it four independent lamp businesses: BTH, Ediswan, Metrovick and Siemens. Metrovick had only a sales organisation; its lamps were branded and made by BTH and Ediswan, who, with Siemens, all had their own factories and sales organisations. BTH was significantly larger, and later that year Chandos took steps to reorganise all these lamp factories and sales organisations into

BTH at Rugby with a new factory at Leicester. Having initiated the process, Chandos moved on to concentrate on more high profile projects in other parts of the company; he also had high profile appointments such as president of the Institute of Directors, all of which probably took his concentration away from the lamps and lighting business. This reorganisation process took five years; during that period it was consolidated under the new name of AEI Lamps and Lighting Ltd. (AEI L&L). However, some bitter internal political battles were fought in the defence of brands scheduled to be dropped; AEI's lamp business suffered badly as all were consolidated under BTH's Mazda brand, which was considered to be the most important. So the famous names of Edison, Swan and Siemens began to disappear, and the combined dominant market share these brands had given AEI L & L was rapidly eroded. The poor management of this process was the gain of GEC, Philips and Thorn, AEI's main competitors, who achieved significant market gains over the same period.

With AEI's total business in a weakened state and its share price slipping, Lord Chandos, the then Chairman, did not see a clear future for the lamps and lighting operations, and Thorn's position had strengthened considerably. This led to discussions between the two companies in which it was decided that neither had the necessary size of production to compete on a world scale. The two companies decided to amalgamate their lighting activities. The first step, in April 1964, was the acquisition of the Thorn lamp and lighting assets by the Thorn lighting subsidiary Atlas Lighting Ltd.; and in June 1964 Atlas Lighting was renamed British Lighting Industries (BLI). Then in anticipation of an amalgamation with AEI L & L, in September 1964 it , for exapmle its capital to provide for the acquisition of Thorn's lamp and lighting assets. The second step, in December 1964, was to further increase the capital for acquiring the issued capital of AEI L & L. These agreements were concluded by the end of December 1964. After the sale agreement, the shares in BLI were 65% held by Thorn, and 35% by AEI. The inter-party agreement covered arrangements for the transfer to BLI of interests in other companies, including Thorn's interest in Ekco. Also, for as long as the shareholding remained in the original split, Thorn had six directors and AEI had three; Thorn had the responsibility for the management of BLI. This merger brought with it a number of changes, which were to influence the industry in the future; some of the key changes were:

- As mentioned earlier in the chapter on 'Components', an important development was the acquisition by BLI from AEI L & L of its interests in the component companies; Glass Bulbs, Lamp Caps, Glass Tubes and Components (GTC) and Lamp Metals, which it now jointly owned with GEC. (AEI and GEC only formally created two of these: GTC and Lamp Metals in 1961.)
- Similarly, BLI acquired the interests in the Controlled Companies (Ismay, Britannia, Splendor, Evenlite and MSL), which it now owned with Philips, GEC and Crompton.
- Interests in the overseas (in Commonwealth countries) lamp companies, which it now held with GEC, Philips and Crompton.
- Interests in a number of wholesalers, particularly 50% in Stern Electric with GEC as the other partner, and 10.5% in Z Electric.
- 40% in British Sealed Beams with Lucas (40%) and GEC (20%) being the other partners. British Sealed Beams was

formed in 1959 by AEI L & L, GEC and Lucas to manufacture the special borosilicate glass used to make sealed beam headlights for motor vehicles and other special lamps, for example Par lamps.

- AEI had a longstanding agreement with Osram GmbH, GE (USA) and Philips for the exchange of patents, technical assistance and know-how. The agreement with Osram GmbH ended in 1966, although BLI did continue to benefit from their patents.

Hence, the creation of BLI had a profound influence on the industry at the time, changing many relationships and creating new ones.

In the 1960s GEC was to undergo major changes. Hugo Hirst and Max Railing had run GEC for over 50 years, with Hugo Hirst as Chairman. When Hugo Hirst died in 1943 at the age of 79, succession had not been structured into the organisation; Harry Railing had been his intended successor, but he died a year earlier. It was said that the death of Hirst's son in 1919, and that of his only grandson and heir, Harold Hugh Hirst, on active service in the Royal Air Force, finished him. In 1943, Harry Railing, the brother of Max, took over as Chairman. This led to 24 years of stagnation and decline, which was not improved when Leslie Gamage, Hirst's son-in-law, took over at the age of 70, in 1957. He put together a management team of long-serving, elderly men, and the GEC financial situation deteriorated to the point where its bankers and financiers started to create pressure for change. Competition from Thorn and Philips in lamps and other smaller companies in other electrical products were eating away at the GEC market share. Consultants were called in, and in 1959 various trading departments were amalgamated into divisions, but it continued to be an unprofitable dinosaur. Gamage left in 1960; Arnold Lindley took over and started to put into action decisions to create a new management structure with a new management team. This was the start of events that would shape the future of GEC for decades to come. In 1961, GEC took over a company called Radio and Allied Industries that was run by a man called Arnold Weinstock; he was immediately taken into Lindley's management team, he took over as Managing Director of GEC in 1963.

By 1960, Philips, who had not taken an aggressive market position, began to realise the growing potential of the self-service retail market. Their position in the wholesale sector had been steadily strengthening, and with it, the growth of fluorescent and discharge lamps. The only major self-service retailer was Woolworths; they dominated this sector and were supplied mainly by the Controlled Companies. Though Philips had a share in the Controlled Companies, it had no influence on this sector of distribution. However, two smaller companies, Kingston Lamps and Luxram Electric, had seen this opportunity and were creating a market niche in supplying the larger supermarkets and cash and carry companies on a direct basis, not through a wholesaler. They also provided a fast delivery service with some in-store merchandising support. In 1964-5, Philips eventually took a controlling share in both these companies to give them an arm length's share and knowledge in this potential growth sector of the domestic market. This was, in later years, to prove a shrewd and rewarding move.

In March 1955, Philips and Thorn came to an understanding with the general intention that Philips should concentrate on GLS and discharge lamps and Thorn on fluorescent lamps. For a limited period, some exchange of products took place, but this was soon to be cancelled by mutual consent in February 1957. In December 1955, Philips and Stella resigned from ELMA because they believed that membership inhibited their commercial development. The competitive discounts offered in the market rose significantly in 1956, creating a price war between the members of ELMA (then comprising AEI and GEC companies, Crompton, Cryselco, BELL and Aurora) on the one hand, and Philips, Thorn and Ekco on the other. Agreed common prices, discounts and rebates continued within ELMA until the Restrictive Trade Practices Act came into force in 1956. In early 1957, ELMA was dissolved, and ELIC (the Electric Lamp Industry Council) was formed with the former members of ELMA plus Thorn, Ekco, Philips and Stella. This new association now included all the larger manufacturers. In an effort to re-establish some stability, ELIC introduced a recommended Non-Mandatory Trading Structure for a period of three months during which Philips and Thorn were to terminate their respective discount arrangements with certain customers. A second structure then was introduced, defining buyers and specifying a compromise discount between the old ELMA and Philips, Thorn and Ekco. The old ELMA discount classification was based upon the buyer's purchases from only ELMA sources. The new ELIC classification was based on the buyer's total purchases of lamps from all sources and the retail prices to which the rates of discount were related were to be determined by the members individually. In practice, they remained almost identical between members. As required by the new Restrictive Trade Practices Act 1956, this trading structure was entered in the register of restrictive trading agreements in September 1957.

It is worth noting that for a few years, a parallel body was formed in 1958: ELIC Ltd. (The Electric Lamp Industry Council Ltd.). This was a vehicle to give financial support to the British Lighting Council, an advisory and promotional body, of which it was one of the founding members. For a time, ELIC and ELIC Ltd., which had the same membership, were operating in parallel. Hence in 1961 it was considered appropriate and more convenient to dissolve ELIC and concentrate all activities in ELIC Ltd.

In June 1959, the Registrar of Restrictive Trading Agreements informed ELIC that he would be issuing a formal notice advising of his intention to refer this registered agreement to the Restrictive Practices Court. Consequently, in November 1959, ELIC informed the Registrar that it had unanimously agreed to abandon the agreement from December 1959. It introduced a discount schedule in January 1960, which, it understood, did not come within the provisions of the Restrictive Trade Practices Act 1956. In March 1960, the Registrar was notified that members of ELIC voluntarily supplied information to one another of changes in discounts to their customers, which had occurred in the previous month. In May 1960, the Registrar was supplied with a copy of the Discount Schedule. While it had similarities, the key difference was that the 1960 Trading Structure did not recommend the rates of discount to be allowed. The purpose of the Schedule was said to be to provide a basis for the collection of information to enable members to determine the discounts to their buyers. The granting of the discount was the sole prerogative of each individual member, as was the definition of any particular class of buyer. Retail list prices, as well as discounts, were fixed by the members of ELIC individually, and were generally the same, type for type, although there were some variations. All the members maintained their respective resale prices of their main brands at all stages of supply. This Trading Structure remained substantially unchanged until 1967.

Other activities of ELIC included the collection and circulation of statistical information that members provided monthly to the Board of Trade. This showed, by quantity, figures of their respective production of all brands of lamps of each of five Board of Trade classifications of electric lamps. Under the same classifications the members also provided, by quantity and value, their home sales deliveries. The members sent copies of these returns to ELIC, where they were collated, and the totals of members' production and home sales deliveries circulated to members. In relation to the introduction of new types, members gave each other six weeks' notice of their introduction. This was to include details of ratings, dimensions and other characteristics.

Unlike ELMA, one of the chief activities of ELIC was in the work undertaken in a number of technical committees and panels of senior technicians who exchanged experiences on design and manufacture. The purpose of these committees was to standardise dimensions and electrical characteristics prior to promulgation of BSI and international standards for lamps. In 1962, the ELIC Council discussed standard incandescent lamps (GLS – General Lighting Service) having a 2500 hour life instead of the normal standard life of 1000 hours. Indeed, a new standard was agreed for this specification; Crompton introduced the lamp under their main brand, promoting it for industrial applications. This discussion on incandescent lamp life continued within the industry and consumers for several decades. Lamp efficiency being derived from light output per wattage versus cost of electricity has been a constant discussion. The relationship between design life, voltage and light output, while being well documented technically, was considered too complex to communicate to consumers. The standard remained both nationally and internationally at 1000 hours. There were certain exceptions; of particular note is Norway, where hydro-electricity was considered so inexpensive the lamp efficiency could be sacrificed for longer life.

A Commercial Committee had been established to undertake activities that included the sponsoring of market surveys, work connected with standardisation of packaging and variety reduction, and co-operation with official bodies and other trade associations. Initially, it was asked to oversee activities such as lists of commercial and industrial users with annual purchases of over £1000, circulating results of local authority tenders and where they had been placed, noting those contracts placed wholly or partly with non-ELIC suppliers. This committee established schedules of multiple retailers, classified according to the number of branches known from members to be selling lamps. The members supplied details of their sales to enable these lists to be classified into 'large retailers' and 'trade users'. It tried, with only very limited success, to establish lists of large users who purchased non-ELIC brands. These numerous activities were to use the ELIC umbrella to gain commercial advantages for members and to regulate the trade. The process of product rationalisation was initiated by this Commercial Committee, and it was to become an activity to be carried out by ELIC and its successor trade association 'The Lighting Industry Federation' (LIF) for many decades to come. Over time the lamp manufacturers had introduced many variations to the basic incandescent lamp, in order to create interest through different shape or finish, variations of light level through different wattages and different caps to match different bulb sizes or to suit imported screw fittings. As no one competitor wanted to allow another to have an advantage, it usually meant that, once a new lamp was introduced, the competitors followed, and this inevitably meant that the catalogues over time had become burdened with many slow-moving products. Strenuous efforts were made over many years to rationalise the product ranges through the Commercial Committee and considerable progress was made; but the process needed to be continuous.

In 1966, the ELIC members agreed that the period of notification of the marketing of new lamp developments would be increased from six weeks to four months. Where field marketing tests were undertaken, members supplied additional information, including the area to be covered, the duration of the test and the number of lamps involved. However, the initiating member could make no public announcement of a new lamp type until four months after the notification date.

In 1968, the Commercial Committee was given the following brief:

> *To consider whether the time was now opportune for the individual members to extend the principle of minimum consignment quantities with the object of more efficient distribution in the best interests of the Trade and general public. At the same time, consideration was to be given to the possibility of further variety reductions having regard to the approaches and published appeals by the Trade for fewer types of lamps than were at present said by retailers to be necessary to hold to give a comprehensive service to the public.*

This study was initially confined to GLS (incandescent) lamps.

By this time, in 1968, the ELIC Commercial Committee had acquired a wide brief for commercial matters including the exchanges of trading information, the classification of buyers, variety reduction, standardisation of packaging, labelling and marking, discussions with other trade associations and government departments on non-technical matters. It also dealt with standardisation of consignment quantities, purchase tax, and new legislation, and lighting promotions with the British Lighting Council (BLC), and the collection of sales records from the three large 'super' wholesalers (Fosters, Stern, Z Electric) of their sales to other wholesalers.

At the same time, the Technical Committee had a defined brief covering the promotion and assistance in the formulation of legislation and codes for improving lamp standards nationally through BSI (British Standards Institute) and internationally through IEC (International Electrotechnical Commission), ISO (International Organisation for Standardisation) and CENEL (Comité European de Co-ordination des Normes Electriques). The Committee members worked closely with the BSI in the preparation of data for lamps for which no BSI specification exists and providing details of lamp interchangeability especially with regard to ratings, dimensions and electrical characteristics.

These two committees provided the backbone of the work carried out by the ELIC organisation.

Another event was also to have a significant influence on the method of carrying out business in the 1960s; this was the abolition of resale price maintenance on lamps. The Resale Prices Act came into force in 1964. ELIC and the EWF (Electrical Wholesale Federation) applied for exemption on behalf of their members who continued to maintain, individually, main brand resale prices on a substantially identical basis. In November 1965, the Registrar of Restrictive Trading Agreements issued a Notice of Reference to the Restrictive Practices Court in respect of electric lamps. After the Registrar's Preliminary Application for Directions in January 1966, ELIC and the EWF duly entered appearances in the proceedings, which resulted in the EWF withdrawing its application. In March, ELIC applied to the Court for an indefinite extension of time for delivery of their statement on the grounds that it would involve them in an undue burden to prepare them to submit to investigation by two different tribunals – the Court and the Monopolies Commission investigation. The Court rejected the application and indicated that it would not refuse to entertain a further application at a later stage, but would be prepared to consider it in relation to circumstances then shown to prevail. Early in 1967, the industry discussed the implications of the end of resale price maintenance; they met with the EWF and agreed a new discount schedule including a major revision of quantity discounts for wholesalers. In March 1967, ELIC announced that it had decided not to proceed further with its application for exemption under the Resale Prices Act 1964, and withdrew from the proceedings. In June 1967, the Restrictive Practices Court refused to make an Order exempting electric lamps from the ban on resale price maintenance.

As a result of the amalgamation of Thorn and AEI L & L in 1967, which was a virtual take-over by Thorn of the lighting interests of AEI, the new company British Lighting Industries (BLI) had more than one third of the market. This resulted in the matter being referred to The Monopolies Commission; the conditions of the Monopolies and Restrictive Practices Act 1948 were deemed to prevail, as BLI now had more than one third of the market. The statement made by the Commission in its summing-up was:[2]

> *The effect on the public interest of the monopoly position of BLI, and therefore of the merger of the lamp and lighting interests of AEI and Thorn which brought it about, is an important issue. However, although BLI now has the largest share of the market, Osram and Philips are powerful companies and the dominance of BLI is far from overwhelming. Indeed, in mercury and sodium discharge lamps alone, apart from other discharge (i.e. fluorescent) lamps, Philips has almost half the trade and BLI has less than one third (although the terms of reference did not permit us to find that the conditions to which the Act applies prevailed in respect of these lamps separately). Apart from BLI's statutory monopoly, therefore, the industry is characterised by oligopoly, and in our inquiry we have been concerned to examine whether, as a result, competition is in any way restricted and, if so, what effect that had on the public interest.*

Market shares in 1967 (following the formation of BLI)[3]

	Filament %	Discharge %	Total %
BLI	39	44	41
Philips	20	31	24
of which: Luxram	3	-	2
Kingston	2	-	1
Osram	23	17	21
Crompton Parkinson	8	6	7
Controlled Companies	6	-	4
Others (incl. BELL)	4	2	3

The full summary of conclusions and recommendations from the 'Second Report on the Supply of Electric Lamps' can be seen in Appendix 2.

By the end of the 1960s the Controlled Companies had been wound up, the industry had undergone radical change and the three large companies of Thorn, Philips and GEC Osram were to dominate the UK market over the next 20 years .

Several industries underwent huge changes in the 1960s and electric lamp manufacturing was no exception. Clearly, the large electrical companies of GEC, English Electric and AEI were very pre-occupied with the heavy electrical side of the industry. This allowed Thorn to concentrate on integrating the lighting interests of AEI; to some extent, the management of GEC was more concerned with the heavy electrical and radio side of their business than Osram, which was becoming more of an arm's length activity but not getting the necessary investment for future activities. While not clear at the time, this was probably the start of the long period during which GEC Osram would make fewer long-term investments, making it vulnerable in the future. It could well have been in Arnold Weinstock's mind that this activity would ultimately be sold back to Osram GmbH, and this did finally happen, 20 years later.

Clearly Thorn now had a dominant position in the lamp and lighting industry. This must have given Sir Jules Thorn great satisfaction as this was always his dream; but the challenge for the future would be to compete on the world market with GE of America, Osram GmbH and Philips. Within the UK, the lamp industry was no longer seen as an exciting progressive industry but as a mature industry from which good financial returns should be made. However, around the corner were events in the form of the oil crisis in 1973 and the power shortage (3-day week) in 1974, which were to shake the lamp industry as never before, throwing it into loss for the first time in its history. This resulted in production rationalisation and closure of some factories;however, it was also anticipated that the period of cheap electricity was at an end, and the larger companies saw it as an opportunity to invest strongly in newer innovative light sources, which would be much more energy effective.

[1] HMSO Report on Supply of Electric Lamps (Monopolies & Restrictive Practices Commission - 1951)
[2] HMSO Second Report on Supply of Electric Lamps (The Monopolies Commission - 1968)
[3] HMSO Second Report on Supply of Electric Lamps (The Monopolies Commission - 1968)

9
The Price of Energy and the Environment

For most of the period since electricity and gas had become widely available during the early part of the 20th century, price had not been a serious issue. Availability, however, was an issue; localised blackouts were not unusual until the 1960s, when much more stability was maintained by a more mature and stabilised electrical supply industry.

Following the Second World War, power generation started to move away from coal with the use of oil as a fuel and the start of nuclear generated power. While air pollution had been a problem for many decades, it was in the period following World War II that solving the issue became a real challenge; there were many contributors to air pollution, but a major one was coal-fired power stations. The use of alternative fuels, in particular oil and emerging nuclear power, were seen as progress. As with all great inventions, many scientists, both in Europe and America, contributed to the invention of nuclear fission; however, the creation of the nuclear reactor has largely been attributed to Enrico Fermi in 1942. He was in charge of the Manhattan Project at the University of Chicago; his team produced the first atomic pile and the first nuclear chain reaction. The nuclear reactor is a system in which a controlled nuclear chain reaction is used to liberate energy, and in a nuclear power plant it is used to generate steam, which operates a turbine and turns into an electrical generator.[1] The first nuclear reactor to produce electricity (albeit in a trivial amount) was the small experimental breeder reactor in Idaho in the USA, which started up in December 1951.[2] The first of several British nuclear plants was Calder Hall 1, which started up in 1956.

This was seen as a new source of cheap and very clean electricity; the issues of decommissioning were not initially fully understood, but there was considerable adverse public reaction to the possible dangers particularly after the Chernobyl disaster. While several countries, notably France and the USA, built a significant number of nuclear plants, Britain remained more cautious and installed more oil and gas fired electrical power generating capacity.

During the Conservative Government of 1970-1974, two major instances occurred which were to influence the cost of power generation for the foreseeable future. In an effort to control inflation during 1972-1973, the Heath Government tried to hold down wages, leading to a confrontation with the trade unions, in particular the NUM, the very powerful miners' union. Initially, they went on a 'work to rule'; later they went on strike.

In the same period, a crisis in the Middle East, that had been rumbling for many years, began to erupt once more. This time, the Arabs were more united in their fight with Israel, and even King Faisal of Saudi Arabia was persuaded by the other Arabian oil-producing countries to put an embargo on crude oil to Western nations in October 1973. The effect of this was felt throughout the world but particularly in the UK, as it coincided with the Government's confrontation with the miners. In order to preserve coal stocks, the Government brought in the infamous 'Three-day week Order' in December 1973, in which industries were shut down for two or three days per week as they had no electricity. Factories with continuous processes, for example glass making, could not manage with intermittent power supply and soon the country was in serious trouble. By 1974, the crisis on *Who governs Britain?* loomed, forcing a general election in February 1974. This resulted in Labour becoming the largest party but with no overall majority. The normal working week resumed in March 1974.

Figure 42: The early Philips compact flourescent SL lamp

This period was particularly turbulent in the UK; but the oil crisis was world wide and caused the end of cheap fuel for ever. The price of oil came down but it was never to be taken for granted again. This became the catalyst for the start of developments for new lower energy lamps giving comparable light output. The first of these, the SL lamp, was launched in the Netherlands by Philips in 1978, but they were quickly followed by an equivalent from Osram. Sometime later Thorn Lighting launched the 2-D lamp, and a new battle was about to start, not just on price but particularly on performance and convenience.

Being a fluorescent lamp, these energy savers are subject to the same technical characteristics as a normal fluorescent lamp, which has always been difficult for many consumers to understand. The light output is not a point source but is approximately evenly distributed over the length of the tube. This light output is normally dependent on the length of the tube; it had similar temperature and starting limitations as a normal fluorescent lamp.

The fluorescent lamp works on the principle of a low-pressure mercury discharge. This tube has an electrode sealed into each

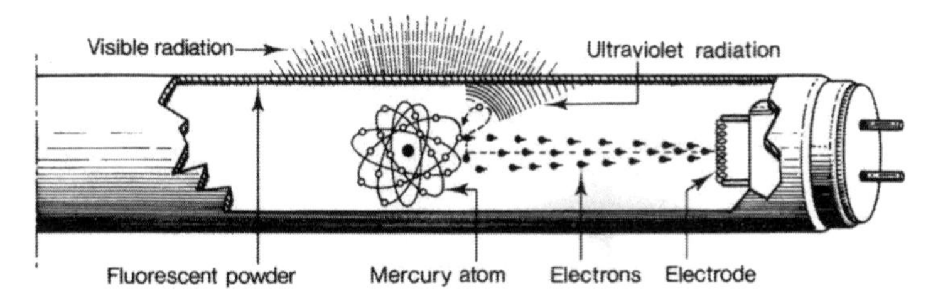

Figure 43: The working principle of a fluorescent lamp[3]

end, and is filled with an inert gas, usually argon, with up to 20%–25% neon and a tiny amount of mercury, which is partly vaporised when the lamp is switched on.

Mercury is used to excite the atoms, and the mercury vapour is very much subject to temperature variation. The chart shown in Figure 44[4] shows the effect of temperature on the light output. This caused problems initially for the users who put these lamps in outside applications during the winter period.

The inside of the tube is coated with a mixture of fluorescent powders which convert the ultra- violet radiation of the mercury discharge into longer wavelengths within the visible range. There are a large number of different fluorescent powders referred to as 'phosphors' mixtures of which produce a variety of different colour temperatures and colour rendering characteristics.

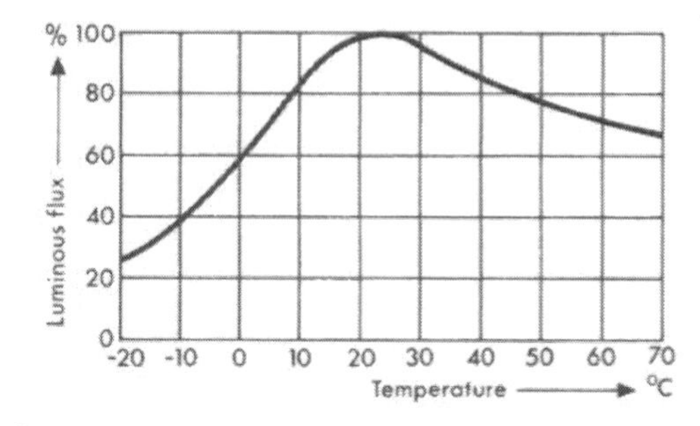

Figure 44: The effect of temperature on light output

To initiate the lamp ignition a simple

circuit is needed; the electrodes are normally preheated, which is achieved by means of a high voltage pulse delivered by a starter switch, and once the arc is formed there must be a current limiting device, called a ballast, which traditionally is a laminated iron core with copper windings, hence the additional weight. These circuit characteristics must also be incorporated in the operation of a compact fluorescent lamp, and therefore the initial lamp developments were heavy and cumbersome. It was this part of the circuit which was the subject of intense development during the 1980s, resulting in electronics achieving the same or similar technical performance, and in a much lighter and smaller lamp circuit. All electronics are susceptible to excess heat, so these had to be carefully placed in the final product in order to avoid these potential excess heat issues.

These lamps are fundamentally fluorescent lamps but they use a much narrower glass tube bent sufficiently to get it into an outer glass envelope, which is often initially rather irreverently referred to as an inverted jam-jar. As these lamps were basically fluorescent, they required both a starter and a ballast; this, like

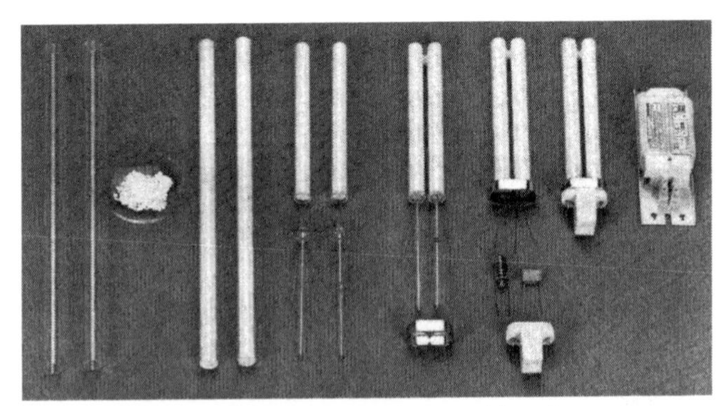

Figure 45: Components of a staged assembly of a compact fluorescent lamp without a glass envelope[6]

most fluorescent ballasts of the time was constructed of laminated iron with a copper winding, which made it very consumer unfriendly and which fitted few domestic situations. This was later to be replaced by electronic ballasts.

Figure 46: Components, including the electronics mounted on a circuit board for the assembly of a compact fluorescent lamp[5]

Their main application in the early phase was in the industrial/commercial market, where their long life (approximately eight times that of an incandescent) was a real benefit; physical appearance was less of an issue. The cost of lamp changes and energy were much more of a priority in the more professional markets than in the domestic market, and the real benefits could be measured. Although energy was to become a major issue, perhaps one of the very major issues the world was to face at that time was that consumers were difficult to convince of the need for this saving. The energy-saving message was a very difficult one to get across. Initially little or no government support was forthcoming for these campaigns, and the sales could not justify long educational campaigns. Hence, during the 1980s, the real marketing campaigns were aimed at the professional markets.

These early lamps were relatively large and heavy. Like most fluorescent and discharge lamps, they took a few minutes to warm up and achieve their full light output. These became negative issues with consumers but again were less important in the commercial and industrial markets, where appearance was generally not an issue; they were mostly placed in applications where long burning hours were the norm, and therefore their initial warm-up period was also not an issue. Furthermore, their weight, due to the heavy copper and iron ballast, became a disadvantage on the BC (B22) cap, as the weight was mainly taken on the two pins; and if it was mounted on its side it was prone to sagging and often even greater weight being taken on one of the pins.

Perhaps most importantly, consumers had not really got the incentive to buy at prices often 20 to 30 times greater than the conventional incandescent lamp. The calculation to demonstrate the cost benefits were too complicated to convey in simple advertisements, and the actual product was not appealing, hence penetration in the early years was very limited.

The next real breakthrough came in the late 1980s with specially designed compact electronics to replace the heavy copper and iron ballast. The electronics reduced the weight and gave other technical advantages of better initial light output and more consistent lumen maintenance; but as with most innovations, these early electronics did not give the reliability needed to give confidence to the users and they were even more expensive that the conventional circuits.

By 1990, official supported agencies were beginning to support the use of these energy saving lamps. Electricity producers and distributors were even showing interest in supporting the lighting industry to gain further penetration into all market sectors, particularly the consumer market. In parallel with these market developments, lower cost production was being developed with manufacturing in both Eastern Europe and the Far East and, in particular, China. This led to more widespread production with many independent Chinese manufacturers producing copies, with subsequent patent infringements. Huge pressure was now being put on the large traditional manufacturers, Philips, GE and Osram, to reduce their prices. Much work went into the redesign of these lamps, and during the 10 years, a large variety of different designs came on to the market. Some of the later types can be seen below:

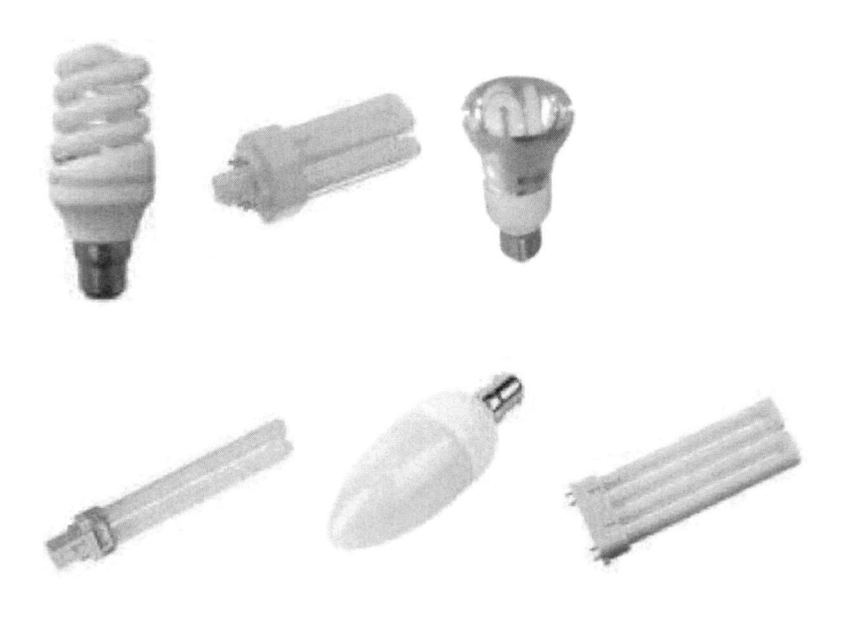

Figure 47: Some of the various shapes and sizes of compact fluorescent lamps (energy savers) on the market

In 1981, Thorn Lighting chose to introduce their own development of the 2-D Lamp, which had both advantages and disadvantages, in the early models, before electronic ballasts were introduced. This design lent itself to being incorporated in many more commercial/industrial luminaries than the initially designed Philips SL or the Osram Dulux. With the strength of the Thorn distribution, particularly in these more professional applications, it gained wide acceptance and was widely specified. Within the first 10 years i.e. the 1980s, this 2-D lamp began to dominate the hotel, restaurant and office sectors in the UK. It was well protected with patents and only supplied by Thorn, which made it one of their most profitable product lines before the Lamp Division was taken over by General Electric (GE) of America in 1991.

The industry in the 1980s continued to develop more sophisticated versions, which were continually being mirrored by the smaller more flexible Chinese manufacturers. Frequently these Chinese versions did not have the comparable quality level; this was often reflected in much shorter life, poorer light level and quality. But the significant factor making them acceptable was the lower price. In addition, new shapes were appearing, many of which had no outer envelope; this made them more vulnerable to breakage but they continued to gain market penetration. This meant that the major Western manufacturers were under serious price pressure, resulting in assembly operations being started in lower-wage countries of Eastern Europe and the Far East, particularly China. Different performance levels were introduced, mainly in the form of differing life time, for example 4000, 8000, 12000, and 16000 hours. Naturally, this led to considerable market confusion but the consumer continued to be mainly concerned with price.

While Chinese competition was initially perceived as having a negative effect on the market, there is little doubt that it stimulated Western manufacturers to speed up their developments and to become more aware of consumer desires, not only for lower prices but also for shapes to fit more domestic lighting fittings. Indeed, the luminaire manufactures also started to design fittings better suited to the new generation of lamps.

The Environment

Almost in parallel with the changes in cost of energy came the serious concern for the environment. The industrialisation of the

Western world was having serious consequences for the atmosphere; without doubt, this was becoming an issue between the First World War and the Second World War, but following the end of World War II, when industry was finally getting back to full output, domestic heating was by coal fires and the atmospheric pollution became so serious that the Government had to act.

While most industrialised areas of the country were affected, it was probably most acutely felt in London. On 5th December 1952, it was cold and there was little wind; fog descended during the day, mixing with large amounts of coal and smoke from the chimneys. Because there was so little wind, and moist ground, the conditions were ideal for radiation fog. It possessed a dry character and when nightfall came, visibility dropped to a few meters. Visibility was very bad for many days. below 50 meters for over 48 hours; at Heathrow airport it remained below 10 meters for almost 48 hours from the morning of 6th December 1952. Road, rail and air transport were brought to a standstill. Theatres had to be suspended when fog in the auditorium made conditions intolerable. However, most importantly, the smoke-laden fog that shrouded the capital brought about the premature death of an estimated 4,000 people who died directly, and a further 8,000 who died later, indirectly

Figure 48: London (December 1952), busses had to be guided through the streets

attributed with many more having serious illnesses attributed to this smog.

This 'pea-soup' smog stayed stewing away for five days from the 5th to the 10th December, as more and more pollution entered it, before winds from the west blew it down the Thames Estuary and out into the North Sea.

The term 'smog' was introduced to describe fog with soot in it; this winter smog had smoke and sulphur dioxide from the City's chimneys accumulating in the foggy air. While in milder form this had been a feature of the London winter atmosphere for almost two centuries since the Industrial Revolution, it was now at totally unacceptable levels. In those few days after 5th December 1952, the air contained thousands of tons of black soot, sticky particles of tar and gaseous sulphur dioxide, which had mostly come from coal burnt in homes. Smoke particles trapped in the fog gave it a yellow black colour. The water from the fog condensed around the soot and tar particles. The sulphur dioxide reacted inside these foggy, sooty droplets to form a solute sulphuric acid, creating in effect a very intense form of acid rain.[7]

The so-called Great London Smog galvanised the Government to pass legislation to clean up the atmosphere, resulting in the Clean Air Act in 1956. This Act was directed at domestic sources of smoke pollution, authorising local councils to set up smokeless zones and make grants to householders to convert their homes from traditional coal fires to heaters fuelled by gas, oil, smokeless coal or electricity. It was not until the 1968 Clean Air Act for Tall Chimneys that industry became the subject of similar legislation. This pollution legislation, plus urban renewal, and the widespread use of central heating in homes and offices, has meant an end to these bad smogs, but pollution continues to cause problems, not just in the UK but all over the world.

Why is all this relevant to the lamp and lighting industry?

Almost since their initial widespread usage, low pressure sodium (SOX) lamps have been disposed of with great care due to their potential fire hazard; local authorities have either directly or, through specialist companies, disposed of these lamps carefully; the inflammable materials have been burnt off and used in landfill sites.

When, during the 1950s and 1960s, shops and offices became widely lit with fluorescent lamps, specialist re-lamping companies became involved; again, these lamps were again disposed of in bulk into landfill sites.

Along with most industries, the lamp industry was not really concerned with the disposal of their products; this was left to users or local authorities. During the 1980s, much greater awareness of waste problems was becoming evident within Germany and several other Central European countries. When the large lamp manufacturers turned their attention to this issue, it became apparent that with care and specialist designed equipment, some lamps, in particular fluorescent, could be broken down into re-usable parts. Certain fluorescent powders, glass, and so on, could be reprocessed and re-used, and hazardous materials disposed of in a controlled manner.

This opened the door to another industry, and while the major lamp manufacturers did not see this as a direct extension of their business, as there was already an industry set up to recover used lamps for disposal, it therefore seemed logical for these companies to be involved in the recycling of these lamps. Initially this industry developed as legislation in Central European countries made it compulsory to remove all hazardous materials for products before being committed to landfill sites. Almost 10 years later, similar legislation was established in the UK.

In 2005, the Government introduced several changes to the Hazardous Waste Regulations, upgrading used fluorescent and sodium lamps (both low pressure - SOX and high pressure -SON); these can release harmful phosphor dust and toxic mercury vapour during disposal. This became part of The Waste Electrical and Electronic Equipment Act 2005, more commonly known as the WEEE directive. This is aimed at encouraging re-using, recycling, and minimising the negative environmental aspects of disposal; the Directive set a target of 65% of IT equipment to be recycled. Putting bulk fluorescent lamps into landfill was no longer an option. It was estimated that, in the UK, 60 million fluorescent lamps were at that time being sent into landfill. It therefore became mandatory for users to have in place adequate systems to manage the disposal of these lamps; this led to specialist companies being set up to manage the process. This legislation was not only directed at manufacturers but also at importers and distributors; this was significant in the electric lamp business, as the major part of the electric lamp business was now imported, not just by the manufacturers but also by independent importers/distributors.

Furthermore, as the use of energy, particularly electrical, within society is increasing, this is a major contributor to atmospheric pollution, in particular the production of carbon dioxide - CO2

As the country headed for the 21st century, the country, indeed much of the developed world, was heading for an energy crisis and serious environmental problems; these were largely caused by excess production of CO2 due to excessive burning of fossil fuels.

Perhaps the lighting industry was not especially large, but its products exclusively used electricity that was rapidly rising in price; producers had the tremendous opportunity to reduce their usage of electricity. This situation was now rapidly being picked up by governments; in California, where there was a critical shortage of electricity generation, the distribution companies started issuing energy-saving lamps (compact fluorescent) to consumers to help to reduce energy usage. Following the Earth Summit in Rio de Janeiro in 2002, the UK Government set a target for the reduction in carbon dioxide by 20% of the 1990 levels by 2010; the Energy Saving Trust was formed to achieve this.

The cost to the environment begins to emerge as a major issue, not just for the UK but, as most would say, for the whole planet. Different countries view this issue from many different perspectives; the developing world sees the issue but does not agree that it should be penalised for problems caused essentially by the developed world, both currently and particularly in their developing past. Issues facing the consumer and business are fundamentally the same: the high cost of energy, and changing attitude on energy efficiency.

Lighting is certainly not the largest user of energy but it is a significant contributor to the overall usage of energy. While it is difficult to be precise, approximately 15% of electricity is used by lighting; this is a little more in the home and a little less in industry. Bigger users of electricity are heating, cooking and industrial processes; and, by the turn of the century, the environment became the major issue and the production of carbon dioxide became the measure for national statistics. This measure brings in usage of gas, oil (petrol/diesel) and other fossil fuels; within this measure, lighting falls to a 4% contribution to the production of carbon dioxide.

Electric lamps provide a very visible source for the use of electricity and, as it was used as a focus in the power crisis of 1974, so it once again very conveniently became the visible focus for energy saving, as the industry had developed energy efficient alternatives for most applications. The Government saw this as an opportunity to once again use the electric lamp as the focus of promotional and public relation campaigns to save energy, hence saving CO_2. This time the lighting Industry was naturally also keen to promote new energy-efficient light sources as the new technology gave rise to new business. Out of the estimated two billion incandescent lamps used in Europe in 2005, approximately one third were in the domestic market and two thirds in industry and commerce.

For many years, the industrial and commercial sectors saw the financial benefits of newer, more efficient light sources, but the most difficult sector was again the domestic market. However, the general public awareness of the need to save energy, and the very visible focus on the electric lamp, created a breakthrough in demand. This was helped by

Daily Mail FREE

Every home will have to use green lighting

EU SWITCHES OFF OUR OLD LIGHTBULBS

Figure 49: Daily Mail 7 March 2007

the industry designing small lamps with higher lumen output and faster ignition; the issue with almost all discharge light sources (including fluorescent) was the time lapse between switch-on and full light output. There was considerable development effort by the lamp industry to improve the electronics used in compact fluorescent lamps for the domestic market.

These modern lamps therefore had very much improved ignition, with almost instant light: 60%–80% light output rising to 100% in 10–20 seconds. This, combined with the lower prices due to the higher volumes and much greater competition, was beginning to make these lamps more acceptable to consumers.

With the industry now being driven not only by competition, but also by legislators and public opinion, huge development efforts were being made to overcome consumer criticisms. At one time the compact fluorescent lamp was seen as the main, if not the only, solution to domestic energy efficient lighting but with the pressure on solving the criticisms, other light sources were reconsidered. Halogen was found to have the potential to meet the efficiency targets and it offered instant start at full light output; and no hazardous materials, for example mercury, a necessary compound in the compact fluorescent lamps, were involved.

With energy now becoming a main issue across the EU, it was not surprising to see this issue becoming the subject of several Directives. A key one was the *Energy Performance of Buildings Directive (EPBD) 2002/91/EC*, to 'Promote the improvement of energy performance of buildings within the Community taking into account outdoor climatic and local conditions, as well as indoor climate requirements and cost-effectiveness'.

Each EU member was then required to transpose this Directive into law by the beginning of 2006, with a further three years being allowed for full implementation of specific articles.

It was estimated that lighting and heating contributes approximately 50% of the total energy consumption and carbon emissions in the UK, of which approximately 15%–18% is from lighting. This meant a challenge to both the industry and consumers alike.

The EU continued its drive to give leadership and legislation to drive greater energy efficiency and lower carbon emissions within buildings. However, within the lighting field, a very significant Directive emerged on 9th March 2007 in which EU Heads of State called for the European Commission to **'rapidly submit proposals to enable increased energy efficiency requirements for incandescent lamps and other forms of lighting in private households by 2009'**. The proposals take the form of implementing measures under the European Union's EUP (Energy Using Products) Directive. The EU also commissioned an independent study by a third-party consultant, Vito (the Belgium Institute for Technological Research). On the 13th April 2009, the Domestic Implementing Measures came into force, as shown in the timetable in Figure 50.[8]

Lamps for special purposes are excluded from these regulations, for example incandescent oven/refrigerator lamps, fireglow lamps and sun tanning lamps.

This table shows the discontinuing of all standard frosted (pearl or coated) incandescent lamps and the gradual phasing out of the standard clear lamps. The death of the standard incandescent lamp is therefore not far away.

EUP ENERGY EFFICIENT - TIMETABLE

Clear Lamps

Stage	*Date*	*Phasing-out*	*Replacements*
1	01-Sep-09	All clear lamps >950 lm (~80w GLS)	Energy class C
2	01-Sep-10	All clear lamps >725 lm (~65w GLS)	Energy class C
3	01-Sep-11	All clear lamps >450 lm (~45w GLS)	Energy class C
4	01-Sep-12	All clear lamps >60 lm (~12w GLS)	Energy class C
5	01-Sep-13	Increased quality requirements	Energy class C
Review	2014		
6	01-Sep-16	All clear lamps >60 lm (~12w GLS)	Energy class B

Non-Clear Lamps

Stage	Date	Phasing-out	Replacements
1	01-Sep-09	All non-clear lamps	Energy class A

Figure 50: Lamp efficiency legislation[9]

The effect of this on the lamp industry was dramatic, and very likely caused the further loss of thousands of jobs throughout Europe. Of course, some would be replaced with the greater demand for energy saving lamps, but by far the most will be in low-wage countries. As with so many industries, electric lamp manufacturing has been almost wiped out in the UK. The skills needed to produce the newer technological products to replace them are no longer available in the UK.

On the positive side, the industry has recognised the lack of acceptance of the fluorescent based energy-saving lamps and has produced mains voltage halogen lamps within typical glass envelopes that look similar to the old standard incandescent lamps and that will also fit into the standard fittings. These lamps are designed to have similar lumen output but with 20%–30 % lower wattages:

28w halogen will replace 40w incandescent
42w halogen will replace 60w incandescent
70w halogen will replace 100w incandescent
105w halogen will replace 150w incandescent

These cost significantly more than the standard incandescent lamp, but will last twice as long and will save approximately 25% of the energy; and they do offer an alternative to the compact fluorescent energy savers, which are not accepted by many consumers.

The table below shows the energy labelling planned to be used with typical domestic lamps:[10]

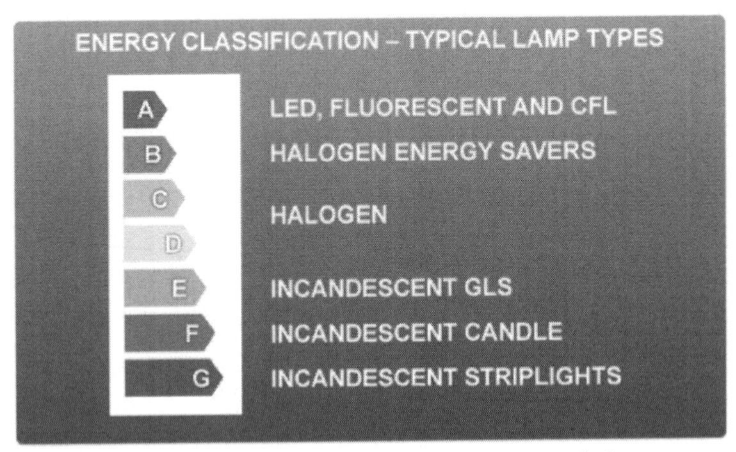

Figure 51: Energy labelling to be used with domestic lamps

The label system below shows the classification of a lamp's electrical energy consumption relative to a standard (GLS or incandescent) lamp producing the same brightness (lumens). The system specified below was the directive drawn up by the European Union in 2011.

Class A & B	Energy savers fall into these categories They are the most efficient type of light bulb and use up to 80% less energy than standard light bulbs
Class D	Mains voltage halogen bulbs usually fall into this category
Class E & F	Standard incandescent light bulbs are the least efficient alternatives.

A	B	C	D	E	F	G
20-50%	50-75%	75-90%	90-100%	100-110%	110-125%	>125%

Lamp class as percentage efficiency of Standard Incandescent light bulbs

However, the future development could be Light Emitting Diodes (LEDs). While this light source has huge potential, much more development is needed to match the requirements of the domestic lighting market. They ignite instantly, are capable of being dimmed, and contain no mercury; potentially, they have efficiency plus a very long life, i.e. up to 50 times that of an incandescent lamp.

Following various international meetings on the environment, the most notable being the 'Kyoto Agreement' in 2001 to reduce future carbon emissions, the UK target was a reduction of 12.5%. The Carbon Emissions Reduction Target supported the large energy suppliers in the reduction of household carbon emissions; and they had overseen over 180 million compact fluorescent lamps (CFL) being mailed out to households by November 2009. Naturally, this was to encourage consumers to use these energy efficient lamps and to prepare the way for the future banning of the standard incandescent lamp. The table below is an example of the kind of savings that could be made by changing lamp use in various applications.

Area of lighting	Lamp type	Energy saving	Replacement Lamp Type	CO^2 savings / lamp / year
Street lighting	HPL	57% ⇨	Ceramic Metal Halide / SON	132 Kg CO^2
Retail lighting	Halogen	80% ⇨	Compact Ceramic Metal Halide	140 Kg CO^2
Office & industrial lighting	T8 fluorescent	61% ⇨	T5 Fluorescent	93 Kg CO^2
Home lighting	GLS incandescent	80% ⇨	CFLi	41 Kg CO^2
LED	GLS incandescent	80% ⇨	LED	41 Kg CO^2

Figure 52: Carbon emissions reduction target

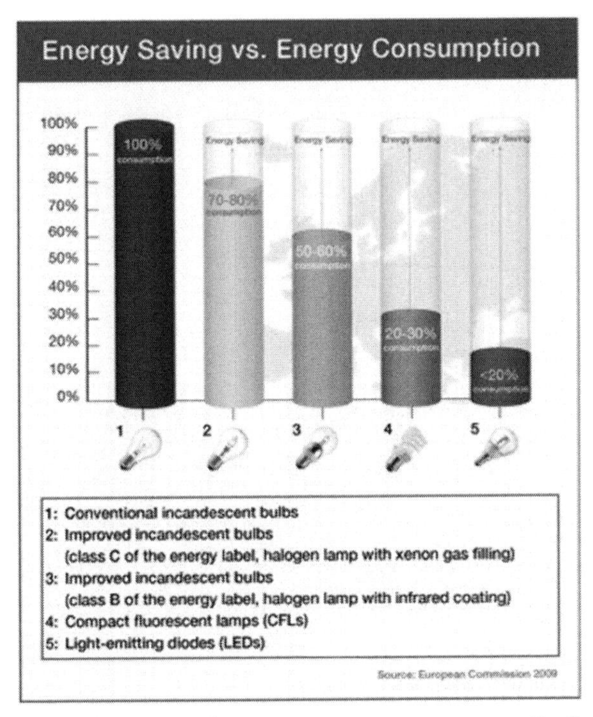

Figure 53: Energy savings versus energy consumption[11]

The electricity supply companies, working with manufacturers and retailers, plan to phase out inefficient light bulbs for domestic use by 2012, saving up to 1.2 metric tons of carbon dioxide by 2020, and around £30 on the average domestic electricity bill:

Coincidental with the new millennium was the real decline in demand for incandescent lamps. The drive for energy saving lamps, and their gradual reduction in cost/selling price, slowly made them more attractive. They found a use in most commercial establishments, for example hotels, offices, and so on. However, in typical consumer applications in the home there continued to be resistance due to their earlier reputation of a flickering start, the time taken to get to full brightness, and the perception they did not give the equivalent light output stated on the packaging. In other words, the real issue with the fluorescent based energy savers was the lack of acceptance.

[1] Paper published on the web by William Erdley of Lancaster University in 2005
[2] Outline of Nuclear Energy by The World Nuclear Association in September 2005
[3] Courtesy of Philips correspondence course booklet no. 9 'Fluorescent Lamps' March 1987
[4] Courtesy of Philips correspondence course booklet no. 9 'Fluorescent Lamps' March 1987
[5] Courtesy of Philips correspondence course booklet no. 9 'Fluorescent Lamps' March 1987
[6] Courtesy of Philips correspondence course booklet no. 9 'Fluorescent Lamps' March 1987
[7] Paper published on the web by The Met Office
[8] Lighting Industry Federation publication
[9] Lighting Industry Federation publication
[10] Lighting Industry Federation publication
[11] European Commission 2009

10
Reaching for the Holy Grail of Light

When the electric incandescent lamp was invented, it was hailed as one of the greatest inventions of the period; indeed, it is often regarded as one of the most life changing inventions of the 19th century. It became the symbol of 'bright ideas' and can still be seen in this respect today, more than 120 years since its invention.

In the post-Second World War period, the industry got a reputation for unacceptable trading practices and even collusion, leading to two Monopolies Commission reports in 1948 and 1967; and consumer and trade confidence in the industry pricing and lamp life began to be eroded.

Initially, this was reflected in a growing pressure on price by the trade and the second Monopolies Commission, recommending the abolition of published recommended retail prices. The manufacturers then began to only publish trade prices on which various different branches of the distribution trade had a series of different discounts. Subsequently, this led to a scramble for higher and higher discounts by the distribution trade, the wholesaler in particular. Wholesalers were beginning to merge and form larger groups with more negotiating power, and consumer lamp sales started to move from the traditional electrical retailer to what was referred to as non-traditional outlets; in the beginning, this was the cash and carry wholesale food distributors.

Woolworths dominated most retail high streets and so were by far the largest retailer of the standard incandescent lamps in 1960s and 1970s. As shown in chapter 6, Woolworths was supplied by Thorn through the Controlled Companies (primarily Thorn and Osram);competitors sought other ways to get to the non-traditional channels, and initially this was the food cash and carry (C & C) outlets. Following the second Monopolies Commission report in 1968, which recommended that the role of the Controlled Companies could no longer be justified, the partners began to wind up this operation. But still the major companies did not want to upset their large electrical wholesale customers; the C & C outlets were supplied by smaller manufacturers, such as Kingston and Luxram, and the second brand operations of the larger companies. Indeed, Luxram acquired British Home Stores (BHS), and it rapidly became their major account, at the same time challenging the Woolworths domination of the consumer market.

Philips could see the changing distribution pattern emerging; they purchased a majority holding in these two companies in the mid-1960s, giving them the opportunity to get a foot-hold and experience in marketing to these outlets. The first major supermarket to enter the lamp business was Sainsbury's, backed by Philips, in the 1970s. As was the Sainsbury tradition, the lamp range was in the Sainsbury brand, and it proved an immediate success. Within the following years in the early 1980s other supermarkets were to follow.

The oil crisis in 1973 saw the end of cheap electricity and the start of the drive for lower energy consumption from all electrical products, and lamps were seen as a very visible sign of electricity consumption. Indeed, the industry was aware that the main domestic incandescent lamp was a very inefficient use of electricity for the production of light – 93% of energy was heat and only 7% light. The major companies, Philips, Osram and Thorn, all initiated research and development programmes with the aim of producing a replacement for this standard incandescent lamp

Philips were the first to launch their SL lamp in 1976; it was referred to as a compact fluorescent lamp (CFL), as it was effectively a miniature fluorescent fitting incorporating a lamp. A miniature fluorescent lamp had been bent double twice; and a starter switch and conventional heavy copper and iron ballast had been all incorporated in a base with ES (Edison Screw) cap and an outer glass put over the lamp (Figure 54).

The wattage had been standardised at 18w, which gave it a similar lumen output to a 75w incandescent lamp. Unfortunately, while this new approach to energy efficient lamps was initially hailed as a very welcome success, it possessed all the worst problems of the fluorescent lamp.

Figure 54: The early compact fluorescent lamp- SL (1976)

It had a switch start, meaning a flickering start, and, as previously explained, a fluorescent lamp is limited to a given number of lumens per unit of its length. Lumen depreciation was rapid in the first 100 hours but it then continued to depreciate less rapidly until failure. It also was larger and had a heavy base due to the ballast (operating gear) in the base, which made it difficult to retro-fit with the standard incandescent lamp. Added to all this was the very high consumer price (approximately £8–£10), so it was not popular in the consumer market.

In reality, the only justification for this was the reduction of energy consumption (it consumed approx. 20%-25% of an equivalent Incandescent lamp) and the longer life of 5000/8000 hours (5-8 times as long). Hence, consumer acceptance of these early versions was very limited, but it did find considerable usage in commercial applications, where it was expensive to change the lamp, and the energy benefits could be easily calculated.

These drawbacks meant that much greater research and development effort was needed to solve these issues. As with so many inventions, the next major development came from the major advances being made in electronics. These enabled these compact fluorescent lamps (CFL) to have an electronic start, creating a faster and non-flickering start; and finally an electronic ballast had much less weight. Of course, this technology was not cheap, and the lamps continued to be expensive.

Figure 55: Electronic circuit on a PL lamp

This electronic circuitry in the early CFL lamps was almost as complex as a radio circuit. As the volume grew and electronic technology advanced, the ability to incorporate parts of the circuit into microchips and the ability to protect the microchips from the heat of the lamp increased; the size of the control gear reduced and reliability improved. With advancing technology and the increase in volume, lower pricing followed. Naturally, this led to a drive for production in low-wage areas, and this inevitably meant a move of manufacturing to Eastern Europe and China. Perhaps production in China also provided a catalyst for lower cost competitors so resulting in a myriad of brands appearing on the market.

While initially the larger manufacturers tried strenuously to protect their patents, competitors found ways around them, and the variety of types appearing on the market multiplied.

As the 21st century came, the drive to save energy gathered momentum and various international agreements were formed to limit carbon emissions. Within this context the energy-saving lamps were being heavily promoted, not just directly by the traditional lamp manufacturers but also by newer entries to the lamp market, by the energy companies themselves in order to meet targets being set for them by governments.

The larger lamp manufacturers were already beginning to regard these, in marketing terms, as mature products. Within the research and development departments, the conceptual use of LEDs (Light Emitting Diodes) for general purpose lighting was beginning to be formed.

The LED was first invented in 1962; this was red, and the first mass product was probably the electronic watch in 1972. By the mid-1970s, several other colours had been developed: yellow, orange and green, but their output was very low. They were used largely by the electronics industry primarily for indicator/signalling lamps, such as illuminating numbers on small screens, for example calculators, and indicator boards; they had been limited to a few colours, primarily red, yellow and green. Initially, these had not been seen as a possible general light source but as their development progressed the lighting industry began to look to LEDs as a real possibility for a light source.

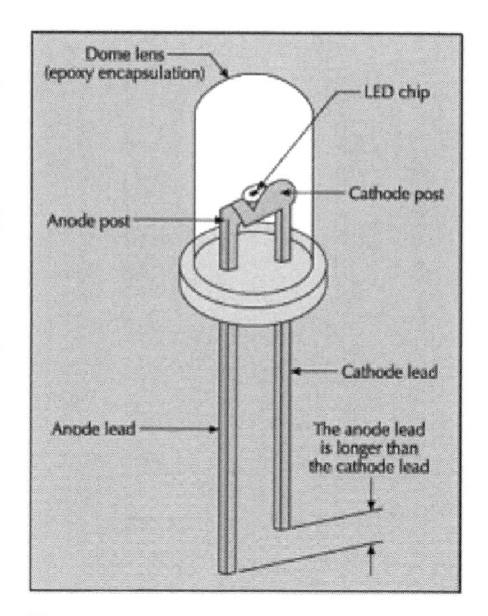

Figure 56: LED technology

The ability to produce a white LED had eluded the electronics industry. However, by the 1980s the first 'super-bright' LEDs were produced, using gallium aluminium arsenide phosphide; the first colours in this were red, then yellow and green. These had an output x30–x40 times greater than the original LEDs.

The rate of development continued; the 1990s saw the race to develop LEDs further, both from their output and in particular, white. Ultra-bright LEDs were produced using Indium gallium aluminium phosphide. The first blue LEDs started to appear. They used silicon carbide, but their output was very dim.

The white was developed in the late 1990s, reputedly by Dr Shuji Nakamura (Nichia Chemical Corporation), the creator of blue, green and white LEDs and the blue laser. He later accepted an appointment to the faculty of the College of Engineering at the University of California at Santa Barbara. Nakamura, known for his technological wizardry with semiconducting gallium nitrides, came to the Santa Barbara faculty from Nichia Chemical Industries in Tokushima, Japan, where he had conducted his research since 1979. He has headed the Department of Research & Development since 1993.[1]

Nakamura had been working on the development of the blue laser; he covered the blue chip with a different phosphor and managed to get the white light, though it was still rather bluish for general lighting applications. The resultant white LED used in a torch shone for 35 hours instead of the equivalent limit for an incandescent torch bulb of six hours.

At the same time, in the mid-1990s, Nakamura was using his blue LEDs to make white LEDs, and adapting his blue LED technology to make a blue laser.

With LEDs, the photons emitted are in a range of similar frequencies, i.e. the blue. With lasers, the frequency of the photons are all the same. To amplify a single frequency of light in a crystal, Nakamura worked out how to etch a fluorescent phosphor on a highly polished mirror on each side of the crystal so that the light bouncing back and forth between the mirrors moved to resonate at the same frequency, thus creating white light. His breakthrough work consisted not only of making the mirrors on the crystal, but also in enabling the crystal to take the high current necessary to create the high-frequency blue laser light.

The blue lasers were substituted for the infra-red lasers used in compact-disc players to get five times as much data on the CD. Blue lasers eventually meant as much as a 35-fold increase in the amount of information that could be contained on a CD, and a major increase in data on a DVD.

All of Nakamura's innovations - the blue, green and white LEDs and the blue laser - depend on the use of the semiconducting material gallium nitride. These research developments based on the material seemed to herald a semiconductor revolution, in which gallium nitride replaced gallium arsenide as the material of choice. Although gallium was common to both materials, the move from its combination with arsenic to the combination with nitrogen was a major innovation in itself. The latter, unlike the former, is environmentally friendly.[2] Intensive research continues at the Engineering Faculty of UC of Santa Barbara in California, and continued improvement can be expected from the LED performance.

The first move in the lighting industry came from Philips, when, in the late 1990s, they entered a joint partnership with Hewlett Packard, a company renowned for the manufacture and development of LEDs for the electronics industry. Recognising the potential for the use of LEDs in the lighting industry, in 1999 they formed a joint company called Lumileds. This was to become the global leader in the production of high-powered LEDs for use in solid-state lighting. In addition, the alternative application that was rapidly developing was for television screens. However, the television industry was seeing the growth of LCD (Liquid Crystal Display) as the main technology for screens in the future. Lumileds was targeted with the responsibility of developing and marketing the world's brightest LEDs. In 2005 Philips acquired the Hewlett Packard shareholding in Lumileds, giving it the leadership in this new emerging technology of solid state lighting.

Naturally, the lighting industry were very quick to see the future for this technology, and again, Philips led the way in trying to acquire the leading developers and register the patents; they were closely followed by Osram GmbH, who, for some years, had been trying to lead the race into new energy-saving light sources. The resultant competitive situation between these two giant European companies has pushed them into being the two global leaders in the lighting field.

This development race is now in full flow; the potential is fully recognised, but there remain many obstacles to these light sources becoming fully accepted. Their life is between 25,000 and 50,000 hours; this is 25-50 times the life of the average incandescent lamp. It comes to full brightness almost immediately, has the capability of fitting into existing sockets, and even has the capability of being dimmed. All these attributes lend this source to becoming a potential replacement lamp in both domestic and many commercial applications. However, as with compact fluorescent lamps, LEDs also have lumen depreciation over their life. With current developments, this is typically 30% after 35,000 hours, but no doubt this will improve with further development. A year has 8760 hours, so this would represent four years of continuous light or 35 years, using the average lit time of 1000 hours per year.

However, it is possible that the real long-term future will be in creating luminaires/light fittings in which the light source is an integral part of the luminaire and not intended to be replaced, because it will last the expected life of the luminaire. This can already be seen in many low voltage applications, for example torches, car lighting, etc. Should the high voltage versions achieve higher output and the potential difficulties with heat generated be adequately dispersed, then this phenomenon could be extended into street lighting and most forms of commercial use, and also into most forms of luminaires mounted in inaccessible locations.

Already many compact fluorescent lamps operated on electronic control gear have a potential lamp life of more than 20,000 hours but still suffer a deteriorating light output however at a much lower rate of deterioration than the earlier lamps. As these are now more mature products, they have achieved a high degree of reliability at lower costs. Hence this light source will continue to provide a competitive alternative in many commercial applications.

While LEDs have been used in the electronics industry for several decades, the white LED only became commercially available at the turn of the 21st century and been used in real lighting applications since then.

Already development of the OLED (Organic Light Emitting Diode) has emerged; this is a completely new concept with possible lighting applications. OLEDs are a solid state device (see Appendix 7) composed of organic molecules that create light with the application of electricity. While their main product application is seen as display screens, including television screens, they can provide crisper displays on electronic devices and they use less power than the current conventional LEDs or LCDs (Liquid Crystal Displays). While the future

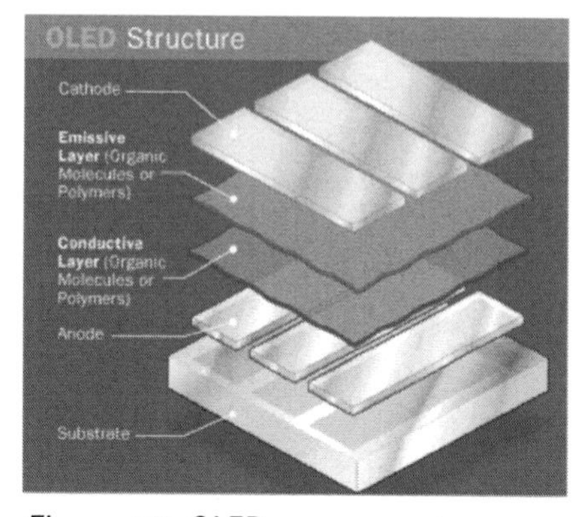

Figure 57: OLED components

application of this technology will probably be for display panels and possibly televisions screens, there is certainly an application for its use in lighting.

As with an LED, an OLED is a solid state semiconductor but it is only 100 to 500 nanometres (1 nanometre = 1×10^{-9} metres) in thickness, or about 200 times thinner than a human hair. OLEDs can have two or three layers of organic material where the third layer helps transport the electrons from the cathode to the emissive layer. The components are illustrated in Figure 57.[3] This technology is more fully explained in Appendix 5.

OLEDs for lighting are also in their infancy, whether they are used as flat or even curved panels of light, or to cover a ceiling, or create a false daylight window. This light source appeals to designers as it lends itself to many new creative lighting concepts. OLEDs, with their special properties, for example ultra low profile, transparency and potential flexibility open up applications such as light tiles, light partition walls and transparent light sources that only emit light after dark, and will act as a window during the day. The greatest successes are expected in applications in which the unique features of OLEDs are used to greatest effect, for example in extremely thin, efficient and transparent two-dimensional light sources with an excellent quality of light.

It is likely to be complementary lighting to be used in conjunction with other more conventional light sources. The success of the development of OLEDs as a light source, particularly from an economic perspective, has yet to be fully demonstrated. This will probably be seen in the next few years, from about 2015. Their future use for display screens is certain, but for lighting it provides a completely new concept and its application is still very much in the conceptual stage.

The future for LEDs is currently seen as general lighting, signalling, including traffic signals, automotive lighting, street lighting, flood lighting, and so on. Development is still in its infancy, however, with white light having been developed in the late 1990s, and types of materials and efficiencies still having many opportunities to improve. In the laboratory, lumen levels of 100 lumens per watt[4] have already been achieved, as have floodlighting schemes with projections of 150 metres..

The real large volume manufacturing has yet to be realised, and this may not happen until some development issues have been solved, such as dissipation of the heat in mains voltage versions, particularly in the higher wattages; also, the higher output in incandescent colour temperatures (colour 3000K), making them as near to fully interchangeable with incandescent lamps as is possible.

In the meantime, there will continue to be a huge advance incorporating LEDs into standard equipment on the basis that they will last the lifetime of that equipment and not be intended to be replaced. While this could extend to consumer lighting, any significant volume has yet to be seen and will take a dramatic shift in consumer perception. However, in the commercial and industrial lighting field, where light is needed in many inaccessible places, and the cost of lamp failure and its replacement far outweigh the cost of the lamp itself, the concept of a light source that will last the life of the lighting installation will be rapidly adopted, for example for street lighting, high mast industrial lighting, etc.

The lamp/lighting market continues to develop at a rapid rate, and particularly now it is linked with the electronics industry, developments are coming through fast. Perhaps the fallacy that the lamp industry has built obsolescence into its products will be completely dispelled in the future.

Philips Lighting, perhaps optimistically, expect LEDs to have 50% penetration by 2015 and 75% by 2020.[5] While this forecast market penetration seems optimistic, the future of the LED as a prime light source is certain.

The old lamp is dead; long live the new lamp.

1 Extract from paper by UC Santa Barbara College of Engineering 1.1.2000
2 Extract from paper by UC Santa Barbara College of Engineering 1.1.2000
3 Extract from paper by Craig Freudenrich Ph.D
4 Philips Lighting (Luimileds)
5 Rudy Provoost, CEO Philips Lighting, April 2010

Appendix 1[1]

Signatories to the Phoebus Agreement

The following parties were signatories to the General Patent and Business Development Agreement dated 20th December 1924 in Zurich, Switzerland, (known as the Phoebus Agreement):

Company	Signatory
'Alpha' Fabrique de Lampes V́ Incandescence, Sà.r.l.	C. Schlesinger
Bergmann Elektrizitäts-Werk A.G.	Steinert, Riehl
British Group:	
British Thompson-Houston Co. Ltd.	P. J. Brewer
Cryselco Ltd.	Wm. F. Moir
Edison Swan Electric Co.	C. H. Cox
General Electric Co. Ltd.	J. Y. Fletcher
Metropolitan-Vickers Electrical Co. Ltd.	R. H. Haviland
Siemens and English Electric Lamp Co. Ltd.	F. Hird
Compagnie des Lampes	M. Saurel
A. C. Cossor Ltd.	W. R. Bullimore
Crompton Parkinson Ltd.	J. Harwood Fryer
Elektra Glühlampenfabrik A.G.	Dr. Ed. Schwartz, Dr. Felix Illner
Aktiebolaget Elektraverken	Henke
Dr. Just Glühlampen U. Elektrizitäts Fabrik A.G.	Aschner
Joh. Kremenezky	Joh. Kremenezky
La Lampe Electrique Belge S. A.	J. Soetewey, A. Bemelmans
Licht A. G. (S.A. Lumière) Goldau	K. Gyr
Manufacture Belge de Lampes Electriques	Pierre Boty

Nordisk Glüdelampe Industri	Rémané, A. Teign
Osram G.m.b.H., Kommanditgesellschaft	Meinhardt, Friedeberg
Osram A.G., Prague	Rosenblatt, Brehmer
Overseas Group:	
Anderson, Meyer & Co. Ltd.	V. Meyer
Australian General Electric. Co. Ltd.	A. Maughan
General Electric Company if Cuba	M. T. McGovern
General Electric, S.A. (Rio de Janiero)	Octavius S. Delaney
General Electric, S.A. (Mexico, D.F.)	M. V. Stewart
Societa Italiana per le Lampade Elettriche "Z"	O. Grauding
South African General Electric Co. Ltd.	R. R. Elliot
N.V. Philips Gloeilampenfabrieken	A. F. Philips
Sté Ame Philips, Eclairage & Radio	Spaens
Julius Pintsch A.G.	Dr. Golen, v.Pabst
Radium Electriztäts G.m.b.H.	E. Kersting
Aktiebolaget Skandinaviska Glödlampfabriken	Almen
Società Edisonper la Fabbricazione della Lampade:	
C. Clerici & C.	O. Grauding
Stella Lamp Co. Ltd.	C. R. B. Leake
Tokyo Electric Co. Ltd.	O. Pruessman
Glühlampen & Elektrizitäts A.G.	Aschner, Fischmann
Vertex Elektrowerk, G.m.b.H.	C. O. Rothweiler
J. Visseaux	J. Visseaux

[1] Second Report on the Supply of Electric Lamps 1968 (The Monopolies Commission) - Appendix 2

Appendix 2

Summary of conclusions and recommendations of the 'Second Report on the Supply of Electric Lamps' by the Monopolies Commission in 1968

The structure of the electric lamp industry has changed in important respects since the Commission previously reported on it in 1951, and in recent years it has become noticeably more competitive to the public interest. The growth of competition and the development of more efficient methods in distribution have been delayed by arrangements and attitudes of mind which have survived from various restrictive practices prevalent in the days of ELMA, but the movement away from traditional practices has accelerated during the course of our inquiry. Although BLI has a monopoly share of the trade defined in the 1948 Act, we have seen the danger to the public interest arising from oligopoly and the main safeguard for the public interest therefore lying in full competition between the leading manufacturers. We have found some restrictions of competition between them still existing, although of declining importance, and we have made recommendations designed to hasten their removal. In other respects we consider that the performance of the electric lamp industry is satisfactory; its technical reputation has always been good, its prices are reasonable by international standards and BLI has recently been successful in export markets, and an improvement in the efficiency of distribution now appears to be in progress.

Our detailed conclusions and recommendations may be summarised as follows:

1. We have found the conditions to which the 1948 Act applies prevail as respects the supply of the lamps covered by the reference.

    - Because at least one third by value is supplied by BLI.

    - Because at least one third is supplied by manufacturers who: (a) recommend or otherwise suggest the retail prices of the bulk of the reference lamps they sell and the distributors generally follow the recommendations. (b) Relate prices for individual buyers to the buyer's total purchases from all suppliers (c) By arrangement exchange information through ELIC about the total purchases of wholesaler and user buyers and about discounts granted to such buyers, and generally inform each other of changes in prices or discounts.

    - Because at least one-third is supplied by members of ELIC who disclose to each other the details of new types of reference lamps in advance of marketing them.

    - Because at least one-third is supplied by BLI and GEC who are joint owners of companies which are the sole manufacturers of certain components for reference lamps cause these manufacturers to discriminate between buyers in the prices charged for the components.

2. The condition which prevail by virtue of BLI's share of the market and by virtue of the exchange of information between the members of ELIC do not operate against the public interest and may be expected not to do so.
3. The conditions, which prevail by virtue of practices, restrictive of price competition and of discrimination in the prices charged for components operate and may be expected to operate against the public interest.
4. By way of remedies for the matters which we have found to operate against the public interest we recommend that:
   - Lamp manufacturers should no longer recommend resale prices for electric lamps, nor should they use lists of retail prices as a basis for trading.
   - In their standard terms lamp manufacturers should no longer base any of their standard discounts on buyers' total purchases from all sources or other assessment of total purchasing potential.
   - All exchange of information between lamp manufacturers concerning prices, discounts and related trading arrangements should cease.
   - Glass Bulbs, Lamp Caps and Lamp Presscaps should charge for supplies of lamp components to their parent companies on the same basis as they charge for supplies to other customers. The range of quantity terms given by these component manufacturers should be only such as can be justified by actual variations in costs. Whatever scales of quantity discounts or rebates they propose to

> apply should be submitted, together with evidence of cost justification to the Board of Trade for approval.

Although we have made no formal recommendations in these respects, we have also expressed the hope that the main brand manufacturers should, when they respond to invitations to public tender, submit genuinely competitive bids more widely than they do at present. That one or more manufacturers should re-examine the desirability of adding a long-life lamp to the range of their lamps offered in the shops. That the membership of ELIC should be extended to all lamp manufacturers, on financial terms they can afford.

Appendix 3

Chronological development of the British Electrical/Lamp Companies

1. British Thompson-Houston (BTH)

1886 American Thompson-Houston took over Laing, Wharton & Down, a London based company, as agents to sell in Britain, apparatus made by the American parent.

1892 The American Thompson-Houston Company merged with Edison General to form General Electric Company (of America).

1894 British Thompson-Houston was formed from the existing Laing, Wharton & Down with additional capital from British and French bankers as an agent for the important GE products into Britain.

1900 Founded Power Securities, as joint undertaking with British Westinghouse, to finance electricity supply companies.

1902 GE appointed Howard C. Levis, Managing Director. (Became Chairman in 1918 and eventually the first Chairman of AEI in 1928.)

1902 Manufacture of electrical equipment commenced at Rugby, with £1million capital plus GE loan of £0.75m. Site comprised 25 acres and produced alternators, steam turbines, motors, converters, switchgear and transformers.

1907	Co-operated with Wolseley Motors to build petrol electric buses.
1909	Supplied electric equipment for the first trolley buses.
1911	Commenced manufacture of tungsten filament lamps under licence from GE of America.
1921	BTH formed a subsidiary Hotpoint Electrical Appliances.
1920's	New factories and extensions built at Rugby, Willesden, Coventry, Birmingham, Chesterfield and Lutterworth.
1921-25	Heavy plant side expanded beyond market capacity and orders taken at low/zero profit.
1926	GE proposed an amalgamation of BTH, GEC, Metropolitan Vickers & English Electric but were rebuffed. They secretly planned to strengthen BTH and purchase through nominees a controlling interest in each of the British companies with a view to forcing a merger.
1926	As part of a campaign to strengthen BTH, GE took over the remaining capital of BTH. BTH took over Ferguson Pailin, and then Ediswan.
1927	The Company, excluding lamps, made a loss of £43,000. GE suspended its more ambitious plans and sought a merger with one of the other British electrical companies.
1927	GE secretly purchased a controlling interest in Metropolitan Vickers (MV).
1927	Sir Guy Granet created Chairman of BTH.

1928 MV on instructions from GE started negotiations with BTH on co-operation in manufacture, engineering and sales.

1928 BTH merged with Metropolitan Vickers to form AEI.

2. British Westinghouse Electric

1899 Company (British Westinghouse Electrical and Manufacturing Co.) founded with £1.25m of which £0.5m was allotted to US Westinghouse for exclusive patent rights. Of the remaining £1.0m capital, the US Westinghouse subscribed £0.5m and the public subscribed the last £0.5m.

1899 George Westinghouse purchased 100 acres of Trafford Park Estate and a further 30 acres. The works were built and fitted out by 1902 and manufacture commenced in 1903.

1904 The first year's results showed a loss of about £110,000 which was concealed with 'optimistic' accounting.

1906 Newcomb Carlton and Philip A. Lang brought in by Westinghouse from parent US Company to resolve financial problems. Carlton instigated the first capital right down of 40% in preference shares and 50% in ordinary stack. He also sold the unused portion of Trafford Park.

1907 US Westinghouse went into receivership due to George Westinghouse mismanagement of the company, which had been making a loss for the past six years. The banks saved the US Company and George Westinghouse was stripped of power in 1910 and died in 1914.

1907 British Westinghouse was saved by Carlton who negotiated a loan of £0.25m but control of the Board passed to the lenders.

1909 The Board of British Westinghouse replaced George Westinghouse as Chairman with John Annan Bryce.

1910 George Westinghouse was voted off the Board. US Westinghouse still held just over half the British Company's capital but the company was now largely independent.

1913 The capital was again written down to £1.15m as compared to the original £3.25m and as a result this and improved business the company made a profit of £100,000.

1913 Decision to purchase US Westinghouse controlling interest in Austrian Lamps.

1914-18 The huge works at Trafford Park were at last put to full use for the war effort making munitions as well as generating equipment.

1917 Lamp manufacture commenced at Trafford Park via Westinghouse interest in Austrian Lamps.

1917 Purchased Brimsdown Lamp Works from the Custodian of Enemy Property.

1918 Dudley Docker reorganised the British Westinghouse Board removing the American Directors.

1918 The Company name was changed to Metropolitan Vickers.

3. Metropolitan Vickers (Metrovick)[1]

The American owned firm of British Westinghouse, formed in 1899 by the American George Westinghouse, was responsible for the formation of Metropolitan-Vickers. British Westinghouse was located at Trafford Park (July 10, 1899 to September 8, 1919) Manchester, an industrial area that became the focal point of many of Metrovick's activities. British Westinghouse felt that the American ownership of its operations during World War One had been a hindrance; thus, in 1916 a British holding company was created to obtain the American shares and the Company became independent from US control. In 1919, the Metropolitan Carriage, Wagon and Finance Company provided the capital for British Westinghouse to become 'British' and British Westinghouse itself was acquired by Vickers Limited. However the company's name changed from 'Vickers Electrical' to 'Metropolitan Vickers Electrical Company' on September 8, 1919.

In 1919, R. S. Hilton became the new Managing Director of Metrovick with G.E. Bailey appointed as works manager. Metrovick was soon faced with the inevitable post-war difficulties; political and labour unrest, import and export restrictions. However, there were positive signs for the company; the tonnage output for 1920 had increased by thirty per cent and the work force to almost 10,000. In 1922, Sir Philip Nash became the new chairman. The early 1920s proved to be a testing time for the newly established firm culminating in the general strike of 1926, which lasted from May 3 to May 12. The Metrovick works survived this period of industrial turmoil with the loss of only one day's work.

1926 witnessed the passing of the Electricity (Supply) Act and the subsequent formation of the Central Electricity Board and the building of an electricity 'grid'. This act was extremely beneficial to Metrovick as the Company was inundated with orders for heavy plant. Metrovick achieved its highest production level in 1927 but this success was overshadowed in January 1928 when Hilton left the company after nine years' service. He was admired by many at Metrovick for his leadership qualities, playing an important part in guiding the company towards being a major competitor in the electrical industry. Hilton subsequently obtained the post of Managing Director of the United Steel Companies, later to be appointed Deputy Chairman, receiving a knighthood in 1942.

The company also expanded into overseas trade when in 1919, the Metropolitan-Vickers Electrical Export Company was formed. The Metrovick overseas trade business accelerated rapidly with offices soon established in Brussels, Bombay, Calcutta, Johannesburg, Melbourne and Sydney. Metrovick was particularly successful in South Africa, Australia and New Zealand. In 1922 a £1 million order to provide traction equipment in South Africa was obtained. The boldest move made by the company was regarding its trading relationship with Russia. The Bolshevik regime was regarded with great suspicion by the outside world and consequently, Metrovick established close relations with the Soviet authorities without government support. Metrovick's association with Soviet Russia was a prosperous one. Important contracts were obtained and in 1924 there were several orders for heavy plant. The 1920s was a period of considerable development for Metrovick with technical advances in the manufacture of turbines, generators, switchgear and industrial motors

In 1928 Metrovick merged with British Thomson Houston (BTH) but on January 4, 1929 both these companies were amalgamated into AEI (Associated Electrical Industries Limited); this resulted in a long history of rivalry between the two firms, which AEI failed to control. The formation of AEI coincided with the Depression of 1929 but there was regeneration in trade from 1933. By 1938 AEI had fully recovered, obtaining Government contracts and increasing its work force to 16,000. The Depression had also affected Metrovick's export business but there were noticeable achievements such the contract for railway electrification in Brazil. Metrovick was also active in Poland and Russia. The effects of the Depression resulted in the departure of Sir Philip Nash in 1931 as Metrovick initiated steps to decrease 'the cost of administration'. Sir Felix J.C. Pole became the new chairman, with the subsequent task of steering the company through the turbulent year of 1933 when six of its engineers were arrested and found guilty in Russia on charges of sabotage and espionage. With the aid of Government intervention, the engineers were released and trade with Russia was resumed after a brief embargo.

The 1930s were years of expansion and innovation for Metrovick but the company was to find its manufacturing capability under considerable pressure during the Second World War. Metrovick, with its site at Trafford Park, was highly involved in the war effort, providing valuable equipment to the British Army, Navy and the RAF. The majority of Metrovick's war production was undertaken at Trafford Park and Sheffield with the Company in close collaboration with various government departments. In 1936, Metrovick was involved in the construction of automatic pilots for the Air Ministry, the development of radar, the manufacture of guns and gun mountings commenced in 1937. Metrovick's greatest achievement lay in the manufacture of bomber aircraft with the company embarking upon an aircraft assembly deal with A. V. Roe in 1938. Metrovick initially began constructing the 'Manchester' bomber before later concentrating on its four-engined successor, the 'Lancaster'. The 1,000 heavy bombers that were produced by the end of the war exemplified success in this field. The culmination of the company's war work was later published in 1947 in the book titled 'Contribution to Victory'.

Metrovick's post-war position was one of increased expansion with the construction of new factories and the internal alteration of existing ones. Captain Oliver Lyttelton (created Lord Chandos in 1954), became Chairman in 1945 with the aim of increasing the efficiency and productivity of AEI. In his first six years as chairman, Lyttelton achieved this objective but was incapable of resolving the commercial rivalry between BTH and Metrovick, which was affecting the stability of AEI. The vast market for generating equipment after the war was extremely lucrative for Metrovick but its competitiveness with BTH intensified. The project at Larne in Northern Ireland, completed in 1957 by BTH, involved the construction of the largest turbine works in Europe. It was hoped that production turbo-generating sets works at Larne would surpass that of the existing factory at Rugby. This Larne factory was resented by Metrovick, who constructed a transformer factory costing £2.5 million, at Wythenshaw. This was a fraction of Larne's £8 million price tag. Despite BTH's new plant at Larne, Metrovick was progressing competitively in the turbine business. Relations between the two rivals again deteriorated when BTH secured the contract for the Buenos Aires power station worth £35 million. Throughout the 1950s Metrovick became established in the manufacture of domestic appliances such as refrigerators and cookers, which became a profitable enterprise for the company.

During his second period as Chairman of AEI (1954-1963) Lord Chandos resolved to extinguish the competition and internal divisionalisation between BTH and Metrovick. The regeneration of AEI was the major goal, which Lord Chandos failed to achieve and his successive attempts at reorganising AEI were ineffectual. It was his struggle to suppress the disorder and conflicting rivalry within AEI, which led to the long-established names of British Thomson-Houston and Metropolitan Vickers being eliminated from the electrical industry on January 1st, 1960. Many Metrovick employees resented this. The remaining years of Chandos' reign were difficult ones for both himself and AEI. In removing the old names, AEI experienced a decrease in profits and share values on the stock market. The indecisiveness of the board and executives at AEI did not assist the company through this turbulent period, as little action was taken to resolve AEI's structural problems. Though Lord Chandos' policy of product divisionalisation was essential to reassert AEI's dominance in the electrical industry, the elimination of the customary names of British Thomson-Houston and Metropolitan-Vickers was deemed unnecessary by those in the engineering world.[2]

1919 Vickers purchased the Docker holdings in Metropolitan Vickers together with the US Westinghouse holding of the equivalent of £1.2m in a very complicated series of financial manoeuvres.

1920 Docker's representatives resigned from the Board of MV and were replaced by Vickers men, Captain Hilton and Major Nash.

1922 Nash succeeded John Annan Bryce as Chairman and managed the business reasonably competently, paying 8% on ordinary and preference shares until 1927.

1924 Nash knighted as Sir Philip Nash.

1927 GE decided to merge BTH with MV and secretly purchased the Vickers interest in MV, registering the shares in the name of Dudley Docker, thus giving the public and MetroVick the impression this was a Docker scheme, in which GE was not involved. Vickers price was £1.3m giving GE 78% of the voting capital.

1928 Bernard Docker, Dudley's son, joined the Board of MV as Deputy Chairman under Nash.

1928 Negotiations started between MV and BTH.

1928 Merger between Metropolitan Vickers and British Thompson Houston is completed and new holding company, AEI was created.

4. Associated Electrical Industries (AEI)[3]

AEI originated in 1929 and began as a financial holding company for a number of leading electrical and manufacturing trading companies in the United Kingdom. These included British Thomson-Houston, Edison Swan and Ferguson Pailin. As the diversity and extent of AEI's products expanded, the Company was joined by Sunvic Controls (1949), Birlec (1954), Siemens Brothers (1955), W.T.Henley (1958) and London Electric Wire Company & Smiths (1959). In 1959 AEI became a trading company and the AEI symbol began to replace the brand names and trademarks of companies within the group (except Lewcos and Birlec).

Sir Felix Pole had been chairman of AEI since its foundation. His years as chairman proved difficult, as he was head of a company that lacked solidarity, especially regarding its activities and the board of directors. One of Pole's primary concerns was the competition between BTH and Metropolitan Vickers. Such rivalry had been present before the merger with AEI in 1928 and was to continue long after. The competition of the 1930's was to significantly affect AEI and measures were taken by the company to reduce the cost of administration. The subsidiary companies of BTH and Metrovick were large exporters during the recession. One notable activity of Metrovic, begun in 1922, was the export of electrical apparatus and machinery to the recently established Soviet regime. This association ended in controversy in 1933 when six Metrovick engineers were tried in Moscow on spying and sabotage charges. Intervention from the British Government resolved the affair and trading with Russia eventually resumed.

If the depression of the 1930s had affected AEI unfavourably, then the Second World War proved economically beneficial for the company. AEI's productive competence was tested as the war progressed. Many factories worked seven-day weeks. The most beneficial aspect for AEI was that it was primarily a war of scientific growth and innovation. The company's electrical engineering products assisted the Government's military projects during the 1930s. Significant contributions to the war effort included automatic pilots for aircraft, radar, guns and gun mountings. However the continuing competition within AEI was underlined by the fact that BTH and Metrovick published separate books detailing their contribution to the war effort. AEI's technical excellence was highlighted in 1935 as Metrovick and BTH became the first two firms in the world to construct jet engines (independently from each other). AEI's greatest work during the war years was its aircraft. In 1938, Metrovick entered a joint venture with A.V.Roe to manufacture aircraft. Metrovick assembled 'Manchester', 'Lancaster' and 'Lincoln' bombers for A.V. Roe at Trafford Park.

At the end of the war in 1945 Sir Felix Pole, now blind, who had been chairman since 1929, was thought to be too old to lead the company into the anticipated post-war boom in electrical equipment. A successor was chosen from outside the company with the resulting appointment of Captain Oliver Lyttelton in the autumn of 1945. His major policy was to reinforce the 'higher direction' of AEI. Uppermost in his agenda was to improve the productivity and 'organise the company along modern lines'. Lyttelton managed to transform the holding company itself into a more proficient organisation during his first six years as chairman (1945-1951). The Conservative victory in the general election of 1951 resulted in Lyttelton receiving the post of Secretary of State for the Colonies. In his absence, Sir George Bailey was appointed chairman. During his three years in charge, Bailey expanded the company's sales and profits; his main achievement was ending the association with GE of America, turning AEI into an entirely British company. In 1954 Oliver Lyttelton returned to AEI as the first Viscount Chandos of Aldershot. Lord Chandos was regarded as an expansionist who was to dominate AEI for a further nine years. He became chairman of the four groups - BTH, Metrovick, Ediswan-Hotpoint and AEI Overseas. Under Lord Chandos the company moved its headquarters to 33 Grosvenor Place, Belgravia, overlooking Buckingham Palace. The most successful achievement of Chandos' second reign was at Larne in Northern Ireland with the completion of a vast works (the largest in Europe) for constructing turbines.

During the mid-1950s AEI was to focus primarily on Domestic Appliances and lighter engineering products. The company discontinued its production of valves and cathode ray tubes and in 1961 merged with Thorn, allowing the latter to manage its interests. AEI purchased Siemens Bros in 1955, thus owning four independent lamp businesses: BTH, Ediswan, Metrovick and Siemens. In subsequently dropping these names AEI's lamp businesses suffered badly. Thus AEI formed a joint company with Thorn again in 1964 and another with EMI in 1966 allowing these companies to manage its domestic appliance businesses. 1950 proved to be a boom year for domestic appliances. Hotpoint, which had been made a separate group in 1955 with Craig Wood as chairman, helped to contribute to AEI's success in this field.

The greatest challenge faced by Lord Chandos, one which had also plagued Sir Felix Pole, was the restructuring of AEI's governing and operational structures. One of Lord Chandos' preoccupations with AEI was his 'divide and rule' strategy. His divisionalisation policy for AEI was designed to mobilise the company's huge assets more effectively and to become more commanding in the markets. After several attempts at revitalisation, Lord Chandos was unable to prevent the eventual unprofitability of AEI and its organisational problems. The years 1960–63 were particularly bad for him and the company, as the serious problem of 'overlapping and competition between the constituent companies was never overcome. Lord Chandos left AEI in March 1963, aged seventy. He had contributed greatly to the company's ascent since the end of the war and, like Hugo Hirst of GEC, believed in the policy of 'Everything Electrical'. The legacy he wished to leave for his successors was one of a 'streamlined company' which would 'survive and prosper in the highly competitive world' which challenged it.

The two men who were to dominate the company until 1967 were Sir Charles ('Mike') Wheeler and Sir Joseph Latham. In 1964 the company's problems were focused upon in a paper entitled 'The State of the Company'. The Wheeler-Latham regime set about altering the hierarchical structure of the company but progress was minimal. During the first two years of the Wheeler-Latham reign, profits were encouraging but it was the disastrous year of 1966 that was to bear more significance. The company had been in need of drastic revitalisation and needed decisive action by the people at the top. In 1967 GEC's Arnold Weinstock and the chairman of the Industrial Reorganisation Corporation, Ronnie Grierson, proposed an instant solution to the company's problems. This was to culminate in the historic £120 million bid by GEC for all AEI, resulting in the merger on Thursday, 9th November 1967.[4]

Chronological Developments of AEI

1929	Sir Felix Pole was appointed Chairman of AEI.
1930	The slump hit AEI badly. Staff salaries were reduced by 5%. Sir Philip Nash was sacked as Managing Director of AEI in 1931.
1932	William C. Lusk appointed Managing Director from BTH and GE.
1930-33	Employees reduced by 1050 to 7750, the salary bill by 17.5% and a further 5% reduction in salaries at the end of 1933.

1935-39	Profits rose sharply as re-armament gathered pace.
1935	BTH accepts a cost plus development contract from Power Jets (Frank Whittle) for the jet engine.
1937	Metrovick commenced development work in conjunction with the Royal Aircraft Establishment on a gas turbine propeller engine.
1938	BTH and John Brown jointly acquire 70% of the Westland Aircraft Company.
1938	Metrovick made aircraft for A.V.Roe (including the famous Lancaster in 1943).
1942-43	BTH and Metrovick withdrew from the jet and gas turbine engine business, which passed to Rolls Royce.
1945	Sir Felix Pole replaced as Chairman by Oliver Lyttelton (later Lord Chandos). He notably failed to stop the overlapping competition between BTH and MV.
1946	GE lowered its holding to 34% under anti-trust pressure.
1951	Oliver Lyttelton joined the Conservative Government and was replaced as Chairman by George Bailey.
1953	GE disposed of the remaining AEI holding.
1954	Lyttelton (as Lord Chandos) returned as Chairman and reigned over AEI for 9 years.

1955	AEI profits after tax £7,863,000.
1955	Work commenced on the new headquarters building in Grosvenor Place (cost £2m).
1955	AEI purchased Siemens Brothers (Lamps) and adopted the BTH trade mark MAZDA on all production and lost substantial business by discontinuing the Siemens, Metrovick, and Ediswan brands.
1956	Chandos set about a very large programme of expansion amounting to £33m.
1955-58	Chandos raised a total of £55m from shareholders.
1957	AEI Board has the first misgivings about the huge capital expenditure.
1957	New transformer factory completed at Wythenshaw for Metrovick.
1957	BTH turbo-generator works completed at Larne, Northern Ireland. It was the largest factory in Europe, the cost £8m and it was never filled.
1958	Profits fell. Chandos dominated the company and the Board was ineffectual. The company was heading for the rocks and nothing was done. BTH and MV continued to compete against each other for business already under pressure from overseas, further driving down price and profits.
1959	Hotpoint became extremely profitable under new management and large sums were invested in expansion just as competition became cut-throat.

	AEI lost confidence and disposed of the business to EMI just as Hotpoint was recovering.
1960	Chandos formally eliminated the BTH and Metrovick names, again losing the goodwill attached to the old names.
1961	Company results were disastrous.
1963	AEI re-organised again. The previous re-organisation was in 1960.
1964	Chandos retires and is succeeded by Sir Charles Wheeler.
1964	Profits rose for the first two years of Wheeler's Chairmanship and then proceeded to decline.
1966	In anticipation of even worse results for 1967, Wheeler was replaced as Chief Executive by Sir Joseph Latham.
1967	Industrial Reorganisation Corporation, set up by the Labour Government in 1966, proposed a merger between AEI and GEC supported by an IRC loan of £15m. Both parties rejected the proposal.
1967	Profits before tax were £3.7m.
1967	GEC makes take-over bid for AEI. The bid terms value AEI at £120m. GEC offered 5 GEC 'B' ordinary shares plus £4 cash for every 8 AEI ordinary shares.
	AEI had to generate cash quickly to provide visible cash to its shareholders. They sold properties including Grosvenor Place (£2.8m) for a total of £4m.

	They merged the telephone exchange business with STC and sold its 35% share of British Lighting Industries to Thorn for £12.3m.
1967 (20/10)	AEI opposed the GEC bid, forecasting increased profits, undertook to take Dr. Beeching on the Board and replace Wheeler with Beeching in 1968.
1967 (30/10)	GEC doubled the cash element in the bid from £4 to £8 for every 8 AEI shares This increased AEI value to £152m.
	AEI rejected the revised bid and GEC raised the bid by 5/- per share increasing AEI value to £160m.
1967 (8/11)	Day after acceptance day, GEC secured over 50% acceptance and AEI passed under GEC control.
1968	GEC, after examining the books, disclosed that AEI had made a loss of £4.5m instead of a profit of £3.7m for 1967.

5. General Electric Company (GEC) – The British GEC

1886 General Electric Apparatus Company

G. Binswanger and Company, an electrical goods wholesaler established in London during the 1880s by a German immigrant named Gustav Binswanger (later Byng), was the building block for GEC.

In 1886 Byng was joined by another German immigrant, Hugo Hirst, (later Lord Hirst) the 'Father of GEC' and the company changed its name to The General Electric Apparatus Company (G. Binswanger). This date is regarded as the real start of GEC.

The following year, the company produced the first electrical catalogue of its kind. In 1888 the firm acquired its first factory in Manchester for the manufacture of telephones, electric bells, ceiling roses and switches.

In 1889 The General Electric Co. Ltd. was formed as a private limited company, also known as G.E.C., with its head office in Queen Victoria Street, London. The company developed the use of china as an insulating material in switches and manufactured light bulbs from 1893. In 1900, GEC was incorporated as a public limited company, The General Electric Company (1900) Ltd.; from 1903 it was styled 'The General Electric Co. Ltd.'

Rapidly growing private and commercial use of electricity, especially in lamps and lighting equipment, ensured buoyant demand and the company expanded both at home and overseas with the establishment of branches in Europe, Japan, Australia, South Africa and India and substantial export trade to South America.

Hugo Hirst had become Managing Director in 1906 and when Gustav Byng died in 1910 he also became Chairman until his death in 1943.

1919 Britain's First Industrial Research Laboratories

During World War I the Company was heavily involved in the war effort with products such as radios, signalling lamps and arc lamp carbons.

In 1919, GEC established Britain's first separate industrial research laboratories at Wembley and moved its head office to new premises in Kingsway, London two years later. From the 1920s the

Company was involved in the creation of the National Grid.

GEC continued to acquire companies and embark on joint ventures, as well as expanding its manufacturing operations overseas and its domestic branch network.

During World War II, GEC was a major supplier to the military of electrical and engineering products. Significant contributions to the war effort included the development of the cavity magnetron for radar, with advances in communications and the mass production of electric lighting.

1961-67 Acquired RAI & AEI

In 1961, GEC took over Radio and Allied Industries (RAI), and with it emerged the new power behind GEC, Arnold Weinstock, who took over as Managing Director in 1963, moving the headquarters of the electrical giant from Kingsway to Stanhope Gate.

Weinstock embarked on a programme that was to rationalise the whole British electrical industry, but began with the rejuvenation of GEC. In a drive for efficiency, Weinstock made cutbacks and mergers, injecting new growth and confidence in GEC, reflected in the profits and financial markets.

In the late 1960s, the electrical industry was revolutionised as GEC acquired Associated Electrical Industries (AEI) in 1967, which encompassed Metropolitan-Vickers, BTH, Edison Swan, Siemens Bros., Hotpoint and W.T. Henley.

1968 Merged with English Electric

In 1968, GEC merged with English Electric, incorporating Elliott Bros., The Marconi Company, Ruston and Hornsby, Stephenson,

Hawthorn & Vulcan Foundry, Willans and Robinson and Dick Kerr.

The background was the rationalisation of the UK heavy electrical industry. The desire of the Central Electricity Generating Board, the principle buyer, was to have only two principal manufacturers for turbo-alternators, the main elements in power stations. A merger of English Electric and GEC-AEI would give 'The General Electric and English Electric Companies Limited' almost exactly half of the turbo-generator business.

On 6th September the two companies issued a joint statement announcing that 'a total merger should be effected between them ... under the chairmanship of Lord Nelson with Arnold Weinstock as managing director'.

GEC continued to expand under Sir Arnold Weinstock, with the acquisition of Yarrow Shipbuilders in 1974 and Avery in 1979.

The late 1980s witnessed further mergers within the electrical industry, with the creation of GPT by GEC and Plessey in 1988, and the joint acquisition of Plessey by GEC and Siemens the following year. An equal investment by GEC and Compagnie General D'Electricitie (CGE) formed the power generation and transport arm, GEC ALSTHOM, in 1989.

The movements towards electronics and modern technology, particularly in the defence sector, showed a marked digression from the domestic market for electrical goods. In 1990, GEC acquired parts of Ferranti and in 1995 acquired Vickers Shipbuilding and Engineering Ltd. (VSEL).

In 1996, Lord Weinstock retired to become Chairman Emeritus after 33 years at the helm of GEC, having become the undisputed leader of the British Electrical Industry.

George Simpson took over as Managing Director of GEC, and with him came a wave of new corporate management. A major reorganisation, aimed at focusing on key business strengths involved the sale of Express Lifts, Satchwell Controls, AB Dick, the Wire and Cables Group, Marconi Instruments and GEC Plessey Semiconductors and the planning of new alliances and acquisitions.

In February 1998, the Company Head Office moved to One Bruton Street, London.

In June 1998, GEC acquired the US defence electronics company TRACOR and it became part of Marconi North America, a Marconi Electronic Systems company.

1999 Acquired RELTEC & FORE

In January 1999 came the announcement of the proposed merger of GEC's Marconi Electronics Systems businesses with British Aerospace. Marconi later disposed of Electronics Systems and sold it to British Aerospace on 29 November 1999, driving Marconi's new focus on communications and IT.

On 1 March 1999 GEC acquired the US telecommunication network products company RELTEC, followed by the announcement of the proposed acquisition of US Internet switching equipment company FORE Systems in June 1999.

The RELTEC acquisition gave GEC, and in turn Marconi products, access to the US market which is where 50 per cent of the telecom's market is. It also reinforced the position in high growth Transmission and Access segment of Communications equipment market.

The acquisition of FORE not only offered a broader range of technology products and a strong enterprise networking clientele, but also strengthened presence in the United States.

1999 Nov GEC Renamed to Marconi plc

On 30 November 1999, GEC renamed to Marconi plc when Marconi was listed on the London Stock Exchange.

This was the culmination of the old GEC's transformation from a holding company of diverse industrial activities to a focused communications and IT company.

In February 2000 Marconi acquired Bosch Public Networks, strengthening Marconi's wireless access product offering and its network management systems.

Chronological Developments of GEC

In the middle of 2001, Marconi's stock dropped to an all time low.[5]

1880 Gustav Byng formed a company G. Binswanger in the City of London and employed Hugo Hirsch to run a department to sell electric lighting accessories.

1883 Hirsch changed his name to Hirst.

1886 The Company name was changed to General Electric Apparatus Company (G. Binswanger), and is joined by Hugo Hirst at 5 Great St. Thomas Apostle.

1887 The first electrical catalogue published by GEC compiled by Hugo Hirst.

1888 The Electric Lighting Act
Factory purchased to manufacture lighting accessories.

1889 Hirst is naturalised.
Company was incorporated with a capital of £60,000 and changed its name to General Electric Company with Gustav Byng as Chairman. GEC moves to 71, Queen Victoria St.
Hirst resigns and is re-instated with 20% of the shares of GEC.

1892 Max Railing comes to England.

1893 Following the expiry of the Edison and Swan lamp patents GEC commenced the manufacture of lamps as Robertson Electric Lamps Ltd.

1894 GEC begins producing Arc lamps.

1895 Robertson Electric Lamps begins the production of carbon filament lamps.
GEC begins producing electric motors.

1898 The Magnet logo is adopted and Magnet House is established.

1900 GEC is floated as a public company – The General Electric Company (1900) Ltd.
The purchase of 42 acres of land at Witton near Birmingham is announced to build a factory to manufacture dynamos, motors and switchgear.

1901 Robertson Electric Lamps went public with a value of £0.5m.
GEC profits were £77,000 for the year 1900 with over 3000 employees.
Hirst's brother-in-law, Max Railing, becomes a Director and joins the GEC board.

1902	The Witton works is opened.
1903	The British General Electric Co. (Pty) Ltd. South Africa The Company becomes The General Electric Co. Ltd.
1906	Hirst appointed sole Managing Director.
1906	Osram Lamp Works set up as joint British Company with GEC and Auergesellschaft to manufacture tungsten lamps at Hammersmith.
1909	Production of Osram tungsten lamps begins.
1910	Gustav Byng dies aged 55 and Hugo Hirst becomes Chairman Dr. A.H. Railing joins the Board. British General Electric Co. (Pty) Ltd, Australia
1912	GEC, BTH and Siemens pool their patents licence and form the Tungsten Lamp Association.
1913	Pirelli-General Cable Works built in Southampton.
1914	Net assets of GEC were £1m and Hirst raised another £0.4m to extend Witton, build Magnet House and buy a half share in Pirelli Cables.
1915	Profits of GEC rose to £138,000. Osram Lamp Works formally purchase Robertson Electric (including Lemington Glass Works)
1916	GEC purchased the Auergesellschaft half of Osram from the Custodian of enemy property for £260,000.
1917	GEC purchased Chamberlain & Hookham.
1918	GEC purchased Fraser & Chalmers of Erith, steam turbine manufacturers, for £600,000.

1919 GEC in conjunction with Marconi's Wireless and Telegraph Co forms the M-O Valve Co at Hammersmith.
English Electric is formed by the amalgamation of Dick, Kerr, Coventry Ordinance Works, and Siemens Dynamo.
Leslie Gamage marries Muriel Hirst (eldest daughter of Hugo Hirst).
Hirst's only son Harold died. Harold's son Harold was born.

1920 Hirst established the GEC research centre at Wembley.
Osram Glass Works opened at Wembley.

1925 Hugo Hirst is made a Baronet.

1929 The Easter Plot
Max Railing becomes joint Managing Director of GEC

1928 GE of America secretly purchased shares in GEC

1929 Over £1m shares in GEC had been bought on the American market, over 50% of the equity. Hirst proposed the issue of £1.6m shares to British subjects only or remove the voting rights of foreign holders.

1929 GE attempts to take over GEC and merge it with AEI. It was unsuccessful and abandoned.

1930 Osram Lamp Works opens at Wembley.

1932 First practical high-pressure mercury discharge lamp invented at Wembley.

1934 Sir Hugo Hirst made Lord Hirst of Witton.

1935 GE shareholding was sold to a British syndicate.

1940 First commercially viable fluorescent tube

1941 Hirst's grandson Harold was killed in the RAF.

1942 Max Railing, Hirst's intended successor died.

1943 Hugo Hirst (79) died leaving no satisfactory successor.

1943 Hirst was succeeded as Chairman by Harry Railing, Max's brother.

1943 Leslie Gamage, Hirst's son-in-law was joint Managing Director and nominated to succeed to the Chairmanship.

1948 The Lamp Agreement

1954 GEC enters nuclear engineering.

1957 Leslie Gamage becomes Chairman.

1959 Osram Glass Works is closed.

1960 Arnold Lindley becomes Chairman and Managing Director.

1961 Arnold Weinstock joins GEC with Radio and Allied Industries.

1963 Arnold Weinstock becomes Managing Director of GEC.
GEC Headquarters established at 1, Stanhope Gate.

1964 Arnold Lindley resigns as Chairman. Lord Aldington takes over.

1967 GEC bid for AEI successful and AEI ceases to exist.

1968 Plessey make take-over bid for English Electric.

1968 Weinstock and Nelson agree merger of English Electric and GEC with English Electric having 29% of the combined Company and 33% of the loan stock.
English Electric ceased to exist.

1968 Lord Aldington stood down as Chairman and was succeeded by Lord Nelson.

1972 GEC took its' option to acquire 50% in British Aircraft Corporation.

1983 Lord Nelson retires. Lord Carrington takes over.

1984 GEC finally sells off TV
Lord Prior takes over as Chairman.

1986 GEC's first offer for Plessey blocked by Monopolies Commission.

1987 Merger of Picker and Philips Medical fails.
Creda and Microscope acquired.

1990 Sale of Osram (GEC) to Osram GmbH (subsidiary of Siemens GmbH)
Osram Lamp Works closed and demolished

1995 Hirst Research Centre at Wembley completely closed and demolished.

1996 Arnold Weinstock retires as Chief Executive and is replaced by George Simpson.

1998 Lord Prior retires as Chairman of GEC and is replaced by Sir Roger Hurn.

1999 Marconi defence business sold to British Aerospace for £7.2 billion.

1999 GEC is reorganised into three divisions:
Marconi Communications
Marconi Systems
GEC Capital
Major concentration on telecommunications market involving huge investments and absorbing the huge cash mountain.

2001 Telecommunications market declines and losses mount.

2002 Marconi becomes insolvent and is involved in major restructuring of debt for equity.

5 History of Siemens AG

German electrical equipment manufacturer formed on Oct. 1, 1966, in the merger of Siemens & Halske AG (founded 1847), Siemens-Schuckertwerke AG (founded 1903), and Siemens-Reiniger-Werke AG (founded 1932). Operating manufacturing outlets in some 35 countries and sales organisations in more than 125 countries, it engages in a wide range of manufacturing and services, with groups for electrical components, computer data systems, power engineering, microwave devices, telegraph and signalling systems, electrical installations, medical engineering, and telecommunications. Headquarters are in Munich.

The first Siemens Company, Telegraphen-Bau-Anstalt Von Siemens & Halske 'Telegraph Construction Firm of Siemens & Halske'), was founded in Berlin on Oct. 1, 1847, by Werner Siemens (1816-92), his cousin Johann Georg Siemens (1805-79), and Johann Georg Halske (1814-90); its purpose was to build telegraph installations and other electrical equipment. It soon began spreading telegraph lines across Germany, establishing in 1855 a branch in St. Petersburg for Russian lines and in 1858 a branch in London for English lines, the latter headed by Werner's brother William (1823-83). As the firm grew and introduced mass production, Halske, who was less inclined toward expansion, withdrew (1867), leaving control of the company to the four Siemens brothers and their descendants.

Meanwhile, the company's activities were enlarging to include dynamos, cables, telephones, electrical power, electric lighting, and other advances of the later Industrial Revolution. In 1890 it became a limited partnership, the senior partners being Carl Siemens (Werner's brother) and Arnold and Wilhelm Siemens (Werner's sons); in 1897 it became a limited-liability company, Siemens & Halske AG.

In 1903 Siemens & Halske transferred its power-engineering activities to a new company, Siemens-Schuckertwerke GmbH (having absorbed a Nürnberg firm, Schuckert & Co.); from 1919 on, the two companies were usually chaired by the same officer, always a member of the Siemens family. In 1932, after seven years of collaboration, an Erlander firm, Reiniger Gebbert & Schall, merged with the Siemens interests to form Siemens-Reiniger-Werke AG, engaged in producing medical diagnostic and therapeutic equipment, especially X-ray machines and electron microscopes.

The House of Siemens, as the companies were collectively called, expanded greatly during the Third Reich (1933-45), all plants running to full capacity during the war and dispersed throughout the country to avoid air strikes in 1943-44. At war's end, Hermann Von Siemens (1885-1986), the head of the group, was briefly interned (1946-48), and Siemens officials were charged with recruiting and employing slave labour from captive nations and associating in the construction and operation of the death camp at Auschwitz and the concentration camp at Buchenwald. As much as 90 per cent of the companies' plants and equipment in the Soviet-occupied zone of Germany was expropriated. The Western powers also removed and destroyed some facilities until the Cold War sparked Western interest in West Germany's economic reconstruction and co-operation. During the 1950s, from its base in West Germany, the House of Siemens gradually expanded its share of the electrical market in Europe and overseas so that by the 1960s it was again one of the world's largest electrical companies. In 1966 all constituent companies were merged into the newly created Siemens AG.

6 History of Siemens & Halske

Werner von Siemens was born on December 13, 1816, in Lenthe, near Hanover, the fourth of 14 children of a less-than-affluent farmer. He quickly developed an interest in science and engineering, but when the family's shortage of money precluded the possibility of a university education on completing school, Werner chose the only viable alternative path in those days technical training as an artillery office with the Prussian army. Once in service, the young officer soon demonstrated that he had special abilities and, according to an appraisal by a superior, was 'thoroughly capable of performing well in the technical field, thanks to his excellent knowledge of engineering and the sciences, and his inventiveness'.

It was during his time with the military that Werner first engaged in business. His brother Wilhelm had filed a patent in England for a method of gold electroplating. The sale of the rights provided the brothers with a sound income for a number of years and allowed Werner to engage in his own research, parallel to his service with the army.

The main focus of his interest was telegraphy, a field that was as yet relatively undeveloped, but Werner nevertheless recognised that it would become a 'technology of the future.' He built a pointer telegraph, an apparatus that proved far superior to similar devices that had been constructed to date. Convinced that his telegraph had the potential to become a success, Werner decided to go into business together with a highly skilled mechanical engineer, Johann Georg Halske, he set up a company, Telegraphen-Bau-Anstalt von Siemens & Halske, in Berlin, which went into business in October 1847.

Inspired by a number of successes that began in 1848 with the construction of a telegraph line between Berlin and Frankfurt/Main, the small company grew so quickly that it soon demanded Werner's undivided attention, prompting him to take his leave of the army in 1849. The entrepreneurial decisions he then made proved pivotal. Since the only contracts for major telegraph lines were likely to come from government offices in those days, it was essential that the company establish a presence abroad. This gave rise to the first branch offices outside Prussia – one in London in 1850 and another in St. Petersburg five years later. The two foreign branches were managed by Werner's brothers Wilhelm and Carl, whom he had involved in the business early on, having taken over the role of head of the family following the untimely death of his parents. Just eight years after starting up, Siemens & Halske had become an international company.

After the first few years of success in business, the scientist and engineer in Werner von Siemens once again came to the fore. In 1866 he achieved his greatest accomplishment: the discovery of the dynamo-electric principle and the invention of a "dynamo-machine", its first practical application. His invention marked the dawn of the age of electrical engineering. (The German term for this field, 'Elektrotechnik', was initially coined by Werner; it had originally been referred to as the 'applied theory of electricity'.) Werner was fully aware of the importance of his discovery: 'Engineering now has the means to produce electric currents of unlimited strength cheaply and easily. This will be of immense significance in many areas within the field as a whole'. And it was with their habitual entrepreneurial vigour that Werner and his company, Siemens & Halske, set about specialising in the areas in question, drives, lighting and power engineering. This was the catalyst that ultimately caused Siemens & Halske to develop into a large-scale enterprise, and by 1890, when Werner retired from the company management; its workforce had grown in number to 5500.

Werner von Siemens, who was raised to the nobility in 1888, also made his mark as a pioneer in a non-technical field and social policy. His view that motivated employees were the basis for the company's success still holds true today: 'It soon became clear to me that the steadily expanding firm could only be made to develop satisfactorily if one could further its interests by ensuring that all employees work together in a cheerful and efficient manner.' He introduced social benefits that were frequently ahead of their time, including a company pension scheme in 1872 (many years before Bismarck introduced national insurance legislation), a nine-hour working day in the same year (when 10-12 hours were the rule elsewhere), and a profit-sharing scheme, the so-called 'stocktaking bonus', launched in 1866.

Werner von Siemens died in Berlin on December 6, 1892. During a full and active life, his interests also extended to public affairs.

As a member of the German Progress Party he held a seat in the Prussian parliament from 1862 to 1866; in 1879 he co-founded the Electrotechnical Society in Berlin; and he set up a foundation to support the Physical-Technical Institute of the Reich, established in 1887.

The name of Johann Georg Siemens has faded into almost total obscurity, but had it not been for him, Siemens, now a major international company, might never even have existed. He was Werner von Siemens' cousin, and it was he who provided the start-up capital of 6,842 thalers needed to launch 'Telegraphen-Bau-Anstalt von Siemens & Halske'. His was the third signature on the articles of association of October 1, 1847, alongside those of Werner and his partner Johann Georg Halske, the company's other two founding fathers. The funds Johann Georg Siemens provided were well invested: The small ten-man operation soon began to thrive. In 1852, just five years after the company was formed, its workforce was 90 strong and its domestic sales exceeded 500,000 marks. Foreign markets were highly important too, even in those days, and export sales ran to almost 450,000 marks. Werner rapidly began to internationalise the company. Foreign branches were set up – first in England (1850), in Russia (1855), and then in Austria (1858). In those days, the workforce in foreign countries exceeded the number of employees in Prussia. Werner showed a keen eye for developing markets and a distinct lack of trepidation about committing himself to business ventures 'at the far end of the world'. The start of operations in Japan, for example, dates back to the early 1860s, and the company installed China's first electric generator in Shanghai in 1879. In 1890, almost half of Siemens' 5,500 employees worked in foreign countries; nine factories generated foreign sales worth 6.6 million marks and domestic sales had risen to 23 million marks. By 1914, Siemens had formed subsidiaries in ten countries and had set up 168 branch offices in a further 49.

In 1897, the family enterprise re-formed as a stock corporation under the name Siemens & Halske AG, a move designed to procure a broader capital base for the company and enable it compete more effectively with a number of strong new rivals, including AEG. Power engineering, which had advanced alongside communications engineering to become the second main pillar of the company's operations, brought sustained growth until World War I. One key event that served to further this development occurred in 1903, when Siemens merged its power engineering activities with the company Elektrizitäts-AG vorm. Schuckert & Co. based in Nuremberg, to form Siemens-Schuckertwerke GmbH (which later became a stock corporation in 1927). Likewise in 1903, Siemens and AEG co-founded Telefunken, which rapidly took the lead in radio and, later, television. In 1913, Siemens achieved sales totalling 400 million marks, and its workforce numbered 63,000.

In 1919, Carl Friedrich, Werner von Siemens' third son followed in the footsteps of his brothers Arnold (1904-1918) and Wilhelm (1918-1919) as head of the company and remained at its helm until his death in 1941 – more than two decades in all. He successfully rebuilt the enterprise, which had lost virtually all of its foreign assets as a result of World War I. Carl Friedrich established a key principle: to concentrate company operations solely on electrical engineering, but also to cover 'the full breadth of electrical engineering.' He re-oriented the company accordingly, withdrawing from 'non-native' areas of business, such as automobile manufacture, and building up other fields instead, such as medical engineering. The latter developed strongly following the 1925 buy-out of Reiniger, Gebbert & Schall, a specialist enterprise for electromedical equipment, based in Erlangen, which was later integrated in 1932 to form Siemens-Reiniger-Werke AG.

The operating results reflect the economic and political instability of the day: In 1923 the Siemens workforce rose above the 100,000 mark for the first time; in 1929 it increased to 138,000. Then, during the Great Depression, 60,000 workers had to be laid off; but by 1939 the payroll had risen to 183,000. Sales fluctuated similarly, rising from 315 million marks in 1924 to 820 million marks in 1929, before dropping to 330 million in 1933. In 1939 the company first posted sales in excess of one billion marks. At the time, Siemens was the world's largest electrical company.

During World War II the company was made to conform to the requirements of the National Socialist wartime economy and was compelled to increase production of goods important to the war effort. The use of forced labour during this era constitutes a dark chapter in Siemens' history.

By the time the war was over, most of the company's plants had either been destroyed or dismantled and roughly 80% of its assets were lost. Siemens began to rebuild at two main locations in the West: Erlangen (Siemens-Schuckertwerke) and Munich (Siemens & Halske). At the end of 1945, Siemens' workforce totalled 38,000. Under the supervision of Hermann (1941-56), Ernst (1956-71) and then Peter von Siemens (1971-81), the company gradually assumed the form it has today. The single most important step in this respect was the merging of Siemens & Halske AG, Siemens-

Schuckertwerke AG and Siemens-Reiniger-Werke AG to form 'Siemens AG, Berlin und München' in 1966. At the time, Siemens was Germany's biggest employer by a wide margin. Its workforce continued to grow, passing the 200,000 mark in 1960 and rising to 300,000 in 1972. Sales increased from DM1 billion in 1951 to more than DM5 billion in 1962, and DM11 billion in 1970.

Earlier, in 1957, Siemens had concentrated its consumer-goods manufacturing in Siemens-Electrogeräte AG which reformed to become a limited liability company. Household appliances passed to the joint venture Bosch-Siemens-Hausgeräte GmbH in 1967. In 1969, Siemens and AEG formed Kraftwerk Union (KWU), which advanced rapidly to become the leading company in the energy sector. In 1977, KWU passed fully into Siemens' hands and has been part of the Power Generation Group since 1987. In 1978, Siemens obtained the full complement of shares in Osram GmbH, a company it had originally co-founded in 1919 with AEG and Auer Gesellschaft.

7 Thorn Lighting Ltd[6]

1928

A man by the name of Jules Thorn established the Electric Lamp Service Company.

1932

Jules Thorn begins making light bulbs in Edmonton, North London. The company grew rapidly to become Thorn Lighting, one of the world's largest producers of lamps, luminaires and lighting components.

1936

The company was strong enough to go public as Thorn Electrical Industries, later to become Thorn EMI.

1948

Thorn is the first company in Europe to mass-produce fluorescent tubes. Thorn research develops a new range of phosphors and improves cathode design.

1954

Thorn introduces the Popular Pack, which becomes the world's best selling fluorescent light fitting.

1967

Thorn research assists development of the high-pressure sodium lamp, ten years ahead of commercial manufacture. The first dip-beam tungsten halogen headlamp for cars helps win the Queen's Award for Technical Innovation.

1972

Thorn is the first manufacturer to offer an integrated system of lighting, heating and ventilation.

1973–79

Acquisition of Kaiser Leuchten, Germany. Thorn's lighting engineering skills combined with a broad product range begin to develop a leading edge reputation.

1981

Thorn introduce the 2D compact fluorescent lamp, an energy-saving replacement for the ordinary light bulb.

1985

Thorn begins manufacture of the high frequency electronic ballast, a major improvement in fluorescent lighting.

1987

Expansion in Europe continues with the acquisition of Jarnkonst, Sweden.

1988

Thorn proves its expertise in lighting management with the launch of the C-VAS system. Acquisition of Europhane in France and ALI/Howard Smith in Australia. Commencement of operations in Hong Kong, Singapore and Malaysia.

1989

Thorn light fittings Aria spotlight and Modulight fluorescent win Die Gute Industrieform design awards at the Hannover Fair.

1991

Disposal of light source manufacturing operation to General Electric of the US. Launch of Sensa, the world's first independent, intelligent light fitting for offices. Acquisition of Philips' airfield lighting business.

2000

Thorn Lighting becomes part of the Austrian Zumtobel Group

[1] Synopsis taken from the Marconi history files
[2] Anatomy of a Merger, by Robert Jones & Oliver Marriott, 1970
'1899-1949 Metropolitan-Vickers Electrical Co. Ltd.', by John Dummelow, 1949
[3] Synopsis taken from the Marconi history files
[4] Anatomy of a Merger, by Robert Jones & Oliver Marriott, 1970
The Memoirs of Lord Chandos by The Bodley Head Ltd., 1962
[5] Information from GEC/Marconi web page
[6] Thorn Lighting web page

Appendix 4

LED Technology[1]

First used as status and indicator lamps within the electronics industry such as illuminating numbers on small screens, for example calculators and indicator boards but limited to a few colours, primarily red, blue and green. The ability to produce a white LED had eluded the electronics industry. Initially these had not been seen as a possible general light source but as their development progressed the lighting industry began to look to LEDs as a real possibility for a light source.

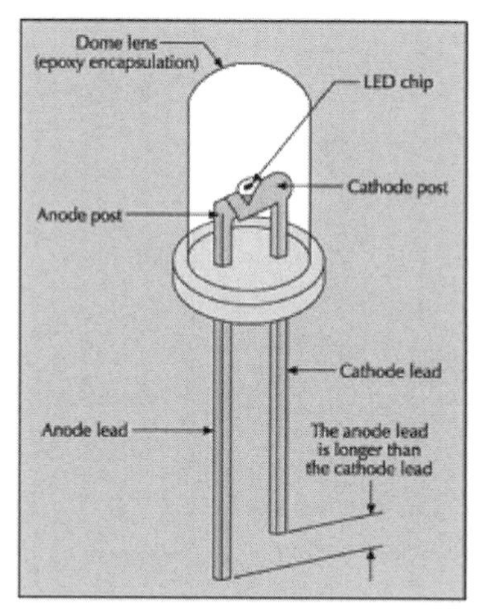

Figure 58: Basic LED components

LEDs are solid-state devices that convert electric energy directly into light of a single colour. Because they employ 'cold' light generation technology, in which most of the energy is delivered in the visible spectrum, LEDs do not waste energy in the form of non-light-producing heat. In comparison, most of the energy in an incandescent lamp is in the infra-red (or non-visible) portion of the spectrum. As a result, both fluorescent and HID lamps produce a great deal of heat. In addition to producing cold light, LEDs:

- Can be powered from a portable battery pack or even a solar array.
- Can be integrated into a control system.
- Are small in size and resistant to vibration and shock.
- Lights up very fast.
- Have good colour resolution and present low, or no, shock hazard.

The centrepiece of a typical LED is a diode that is chip-mounted in a reflector cup and held in place by a mild steel lead frame connected to a pair of electrical wires. The entire arrangement is then encapsulated in epoxy. The diode chip is generally about 0.25 mm square. When current flows across the junction of two different materials, light is produced from within the solid crystal chip. The shape, or width, of the emitted light beam is determined by a variety of factors: the shape of the reflector cup, the size of the LED chip, the shape of the epoxy lens and the distance between the LED chip and the epoxy lens. The composition of the materials determines the wavelength and colour of light. In addition to visible wavelengths, LEDs are also available in infra-red wavelengths, from 830 nm to 940 nm.

The definition of 'life' varies from industry to industry. The useful life for a semiconductor is defined as the calculated time for the light level to decline to 50% of its original value. For the lighting industry, the average life of a particular lamp type is the point where 50% of the lamps in a representative group have burned out. The life of an LED depends on its packaging configuration, drive current and operating environment. A high ambient temperature greatly shortens an LED's life.

Additionally, LEDs now cover the entire light spectrum, including red, orange, yellow, green, blue, and white. Although coloured light is useful for more creative installations, white light remains the holy grail of LED technology. Until a true white is possible, researchers have developed three ways to deliver it:

- Blend the beams. This technique involves mixing the light from multiple single-colour devices. (Typically these are red, blue, and green.) Adjusting the beams' relative intensity yields the desired colour.
- Provide a phosphor coating. When energised photons from a blue LED strike a phosphor coating, it will emit light as a mixture of wavelengths to produce a white colour.
- Create a light sandwich. Blue light from one LED device elicits orange light from an adjacent layer of a different material. The complementary colours mix to produce white. Of the three methods, the phosphor approach appears to be the most promising technology.

Another shortcoming of early LED designs was light output, so researchers have been working on several methods for increasing lumens per watt. A new 'doping'technique increases light output several times over compared to earlier generations of LEDs. Other methods under development include:

- Producing larger semiconductors.
- Passing larger currents with better heat extraction.
- Designing a different shape for the device.
- Improving light conversion efficiency.
- Packaging several LEDs within a single epoxy dome.

One family of LEDs may already be closer to improved light output. Devices with enlarged chips produce more light while maintaining proper heat and current management. These advances allow the units to generate 10 times to 20 times more light than standard indicator lights, making them a practical illumination source for lighting fittings.

Before LEDs can enter the general illumination market, designers and advocates of the technology must overcome several problems, including the usual obstacles to mainstream market adoption: Industry-accepted standards must be developed and costs must be reduced. But more specific issues remain. Things like lumen-per-watt efficacy and colour consistency must be improved, and reliability and lumen maintenance should be addressed. Nevertheless, LEDs are well on their way to becoming a viable lighting alternative.

1 Understanding LED Technology' by Joe Knisley, Senior Editorial Consultant 2002

Appendix 5

OLED Technology[1]

In contrast to traditional point-like inorganic light emitting diodes (LEDs) that are based on complex crystalline structures, OLEDs are flat light sources that use organic semiconductors to generate light. They consist of a (glass) substrate, a transparent electrode, one or more thin organic layers, and a counter-electrode, which can also be transparent. The component is encapsulated to protect against oxidation and moisture. An OLED operates on the principle of injection electro-luminescence, just like an inorganic LED. Positive and negative charge carriers are made to give off light by recombining in an emission layer. In other words, light is emitted in a semiconductor – an inorganic semiconductor in the case of an LED and an organic one in the case of an OLED. The structure of the molecules determines the colour of the light; in the case of LEDs the light colour is determined by the crystal structure of the semiconductor materials

Like an LED, an OLED is a solid state semiconductor but is only 100 to 500 nanometres

(1 nanometre = 1×10^{-9} metres) in thickness or about 200 times thinner than a human hair. OLEDs can have two or three layers of organic material where the third layer helps transport the electrons from the cathode to the emissive layer. The components are illustrated in Figure 59.

These consist of:-

- **Substrate** (clear plastic, glass, foil) - The substrate supports the OLED.
- **Anode** (transparent) - The anode removes electrons (adds electron 'holes') when a current flows through the device.
- **Organic layers** - These layers are made of organic molecules or polymers.
- **Conducting layer** - This layer is made of organic plastic molecules that transport 'holes' from the anode. One conducting polymer used in OLEDs is polyaniline.
- **Emissive layer** - This layer is made of organic plastic molecules (different ones from the conducting layer) that transport electrons from the cathode; this is where light is made. One polymer used in the emissive layer is polyfluorene.
- **Cathode** (may or may not be transparent depending on the

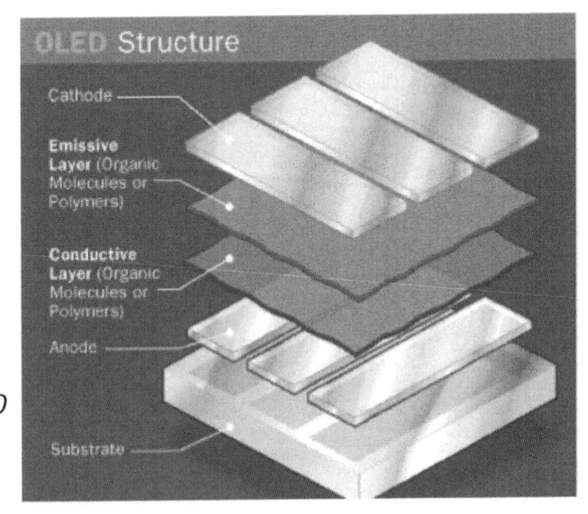

Figure 59: OLED components

type of OLED) - The cathode injects electrons when a current flows through the device.

The working principle of the OLED

OLEDs emit light in a similar manner to LEDs, through a process called electrophosphorescence. When a power supply is connected to a device containing the OLED applies a voltage across the OLED then:

1. An electrical current flows from the cathode to the anode through the organic layers (an electrical current is a flow of electrons).
   - The cathode gives electrons to the emissive layer of organic molecules.
   - The anode removes the electrons from the conductive layer of organic molecules. (This is the equivalent to giving electron holes to the conductive layer.)

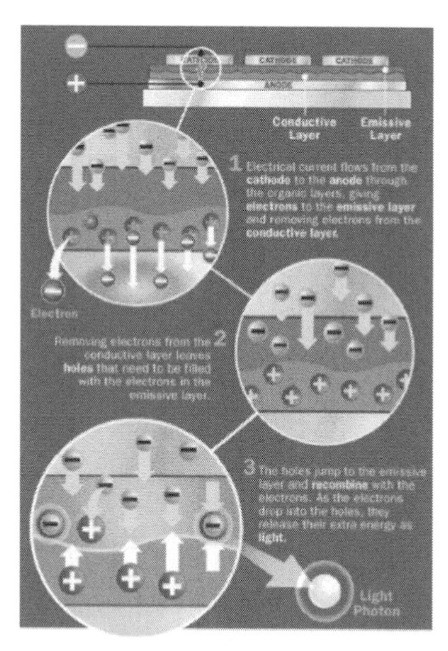

Figure 60: The working principle of the OLED in creating light

2. At the boundary between the emissive and the conductive layers, electrons find electron holes.
    - When an electron finds an electron hole, the electron fills the hole (it falls into an energy level of the atom that's missing an electron).
3. When this happens, the electron gives up energy in the form of a photon of light.
4. The OLED emits light.
5. The colour of the light depends on the type of organic molecule in the emissive layer. Manufacturers place several types of organic films on the same OLED to make colour displays.
6. The intensity or brightness of the light depends on the amount of electrical current applied: the more current, the brighter the light.

There are several types of OLEDs:

- Passive-matrix OLED
- Active-matrix OLED
- Transparent OLED
- Top-emitting OLED
- Foldable OLED
- White OLED

White OLED

White OLEDs emit white light that is brighter, more uniform and more energy efficient than that emitted by fluorescent lights; white OLEDs currently have a good colour rendering index (CRI ≥

80) and potentially have the true-colour qualities of incandescent lighting. It is expected that they will be capable of being dimmed by adjusting the current and, unlike fluorescent lamps, give full output as soon as they are switched on. OLEDs can be made in large sheets; they can replace fluorescent lights that are currently used in homes and buildings.

It is a two-dimensional light source that provides a soft light that is largely glare-free with no harsh shadows. It is possible to look directly at OLED light sources without being dazzled – in complete contrast to conventional light sources such as incandescent lamps, halogen lamps and LEDs. OLEDs are extremely thin. Their active layers have an overall thickness of less than 500 nm, which is 100 times thinner than a human hair. The total thickness of the component is typically 1.8 mm, comprising the glass substrate and the encapsulation; and even this can be reduced considerably by using thinner and more flexible substrates and thin-film encapsulation. OLED panels therefore weigh very little indeed.

OLEDs are more than just light sources, they are design elements. Even when they are switched off they look very different from conventional light sources. They are very thin, very flat, very lightweight and therefore very attractive. They can also be transparent, diffused or mirrored so they offer a completely new look. Luminaire designers can therefore look forward to enormous freedom.

OLED light sources could eventually replace some volume business in the general lighting sector. The mass-market phase will start as soon as OLEDs can be manufactured in very high quantities and with consistently high quality and when appropriate demand has built up on the world market. As economical and ecological light sources, OLEDs will enrich the broad spectrum of general lighting products and make attractive additions to the lighting portfolio. This is not likely to be seen until at least 2015.

[1] Extract from paper by Craig Freudenrich Ph.D

Bibliography

Archives

The British Library

The Institute of Electrical Engineers Archives

Marconi / GEC Archives

Bibliography

The History of N. V. Philips Vol. 1 & 2 by A. Heerding (Translated by Derek S. Jordan)
First published in English by Cambridge University Press in 1988

The History of N. V. Philips Vol. 3 & 4 by I. J. Blanken (Translated by C. Pettiward)
Published by European Library – Zaltbommel, The Netherlands (printed in English in1999)

Lengthening the Day – A history of lighting technology by B. Bowers
Published by Oxford University Press in 1998

Anatomy of a Merger – A history of GEC, AEI and English Electric by Robert Jones & Oliver Marriott
Published by Jonathan Cape Ltd., London in 1970

The Electric Lamp Industry – Technological change & economic development from 1800-1847, by A. A. Bright, New York 1949

From the Beginning – Robertson Electric lamps by H. Loring
Published by Jarrold & Sons, London 1905

The Rise and Fall of the UK Electrical Industry 1882-1999 by
Frederick B. Shorrocks

The GEC Research Laboratories 1919-1984 by Sir Robert Clayton
& Joan Algar
Published by Peter Peregrinus Ltd in association with the
Science Museum, London.

BTH Reminiscences – Sixty Years of Progress by H. A. Price-
Hughes
Published by British Thompson – Houston Co. Ltd.

Sir Joseph Wilson Swan FRS – Inventor and Scientist by Mary E.
Swan and Kenneth R. Swan
Published by Oriel Press, London 1929

Siemens Brothers 1858-1958, by J. D. Scott
Weidenfeld and Nicolson, London 1958

R. E. Crompton – Reminiscences
Published by Constable & Co Ltd. in 1928

Sir William Siemens – A man of Vision

The Story of the Lamp

Published by The General Electric Co. Ltd. London 1924

The GEC, Its History, Structure and Future by T. W. Heather
Published in London 1953

Lamps and Lighting by H. Hewitt & A. S. Vause
Published by Edward Arnold (Publishers) Ltd., 1966

High Pressure Mercury Vapour Lamps and their Applications by W.

Elenbaas
Published by Macmillan & Co. Ltd., London 1965

Miscellaneous References

Letter from Lane Fox Electrical Company Limited on 'The Lane Fox Patents', dated 31st May 1890, sent by C. A. Stephenson, Managing Director

Various editions of 'The Electrician' published during the 1880s

'On the Electric Light of Mercury' an article by J.H.Gladstone in Philosophical Magazine

Faraday – Plücker correspondence in the archives of the Institution of Electrical Engineers, collection SC MSS 2, D. J. Blaikey papers

Paper presented by Joseph Swan to the Literary and Philosophical Society of Newcastle on 3.2.1879

Paper presented by Joseph Swan to the Institution of Electrical Engineers in London, from The Engineer, 29.10.1880

Report from the Standing Committee on Lighting by Electricity, 1879

Findings & Decisions of the sub-committee appointed by the Standing Committee on Trusts, 1920

History of Light and Lighting by G.W. Stoer - Published by Philips Lighting

UV Radiators and Applications by Philips Lighting

IR Radiators and Applications by Philips Lighting

Paper on 'Lighting A Revolution' from the National Museum of American history

HMSO Report on the Supply of Electric Lamps (Monopolies & Restrictive Practices Commission - 1951)

HMSO Second Report on the Supply of Electric Lamps-1968

Various documentation published by the Lighting Industry Federation, the Lighting Association, and The European Commission